Wastewater Treatment

Occurrence and Fate of Polycyclic Aromatic Hydrocarbons (PAHs)

Advances in Water and Wastewater Transport and Treatment

A SERIES

Series Editor
Amy J. Forsgren

Xylem, Sweden

Wastewater Treatment: Occurrence and Fate of Polycyclic
Aromatic Hydrocarbons (PAHs)
Amy J. Forsgren

Harmful Algae Blooms in Drinking Water: Removal of
Cyanobacterial Cells and Toxins
Harold W. Walker

ADDITIONAL VOLUMES IN PREPARATION

Wastewater Treatment

Occurrence and Fate of Polycyclic Aromatic Hydrocarbons (PAHs)

Edited by

Amy J. Forsgren

CRC Press
Taylor & Francis Group
Boca Raton London New York

CRC Press is an imprint of the
Taylor & Francis Group, an **informa** business

CRC Press
Taylor & Francis Group
6000 Broken Sound Parkway NW, Suite 300
Boca Raton, FL 33487-2742

First issued in paperback 2017

© 2015 by Taylor & Francis Group, LLC
CRC Press is an imprint of Taylor & Francis Group, an Informa business

No claim to original U.S. Government works

ISBN-13: 978-1-4822-4317-8 (hbk)
ISBN-13: 978-1-138-89390-0 (pbk)

Library of Congress Cataloging-in-Publication Data

Wastewater treatment : occurrence and fate of polycyclic aromatic hydrocarbons (PAHs) / editor Amy J. Forsgren.
    pages cm. -- (Advances in water and wastewater transport and treatment ; 2)
    Includes bibliographical references and index.
    ISBN 978-1-4822-4317-8 (alk. paper)
    1. Sewage--Purification--Organic compounds removal. 2. Polycyclic compounds--Environmental aspects. I. Forsgren, Amy.

TD758.5.O75W37 2015
628.3--dc23
2014043709

Visit the Taylor & Francis Web site at
http://www.taylorandfrancis.com

and the CRC Press Web site at
http://www.crcpress.com

To Rachel Carson, whose book Silent Spring *was a major driving force behind the creation of the USEPA, and to Frances Oldham Kelsey, the scientist who put a human face on the words teratogen and mutagen*

Contents

Contributors

David Bolzonella
Department of Biotechnology
University of Verona
Verona, Italy

Andres Campiglia
Department of Chemistry
University of Central Florida
Orlando, Florida

Jared Church
Department of Civil,
  Environmental, and
  Construction Engineering
University of Central Florida
Orlando, Florida

Shirley Clark
Penn State Harrisburg
Middletown, Pennsylvania

Francesco Fatone
Department of Biotechnology
University of Verona
Verona, Italy

Amy J. Forsgren
Xylem Inc.
Sundbyberg, Sweden

Kenya L. Goodson
Nspiregreen, LLC
Washington, DC

Xuefei Guo
Department of Chemistry
University of Cincinnati
Cincinnati, Ohio

Imen Haddaoui
Higher Institute of Agronomic
  Sciences of Chatt Meriem
Tunis, Tunisia

Woo Hyoung Lee
Department of Civil, Environmental,
  and Construction Engineering
University of Central Florida
Orlando, Florida

Aleksandra Jelic
Department of Biotechnology
University of Verona
Verona, Italy

Evina Katsou
Department of Biotechnology
University of Verona
Verona, Italy

Jan Kochany
Environmental Consultant
Mississauga, Ontario, Canada

Xiangmeng Ma
Department of Civil,
  Environmental, and
  Construction Engineering
University of Central Florida
Orlando, Florida

Olfa Mahjoub
National Research Institute for
  Rural Engineering, Water, and
  Forestry (INRGREF)
Tunis, Tunisia

Simos Malamis
Department of Biotechnology
University of Verona
Verona, Italy

Daniel Mamais
Department of Water Resources and
    Environmental Engineering
School of Civil Engineering
National Technical University of
    Athens
Athens, Greece

Robert Pitt
Department of Civil, Construction,
    and Environmental Engineering
University of Alabama
Tuscaloosa, Alabama

Agnieszka Popenda
Department of Chemistry
Water and Wastewater Technology
Częstochowa University of
    Technology
Częstochowa, Poland

Vincenzo Torretta
Università degli Studi dell'Insubria
Varese, Italy

Maria Włodarczyk-Makuła
Department of Chemistry
Water and Wastewater Technology
Częstochowa University of
    Technology
Częstochowa, Poland

Daoli Zhao
Department of Chemistry
University of Cincinnati
Cincinnati, Ohio

Acronyms

ATSDR	U.S. Agency for Toxic Substances and Diseases Registry
BaP, B(α)P	Benzo[a]pyrene, also known as benzo[α]pyrene
BOD_5	5-day biochemical oxygen demand
BSA	Bovine serum albumin
BTEX	Benzene, toluene, ethylbenzene, xylene
CAS	Chemical Abstract Services
CASP	Conventional activated sludge process
$CBOD_5$	5-day carbonaceous biochemical oxygen demand
CNT	Carbon nanotube
COD	Chemical oxygen demand
CT	Coal tar
DMSO	Dimethyl sulfoxide
DNA	Deoxyribonucleic acid
DO	Dissolved oxygen
dw, d.w.	Dry weight
EC	European Commission (EU's executive body)
EEA	European Environmental Agency
EEM	Excitation-emission matrix
EF	Emission factor
EFSA	European Food Safety Authority
E.I.	Equivalent inhabitant
ELISA	Enzyme-linked immunosorbent assay
EPA	*See* USEPA
EQS	Environmental Quality Standards (European Union)
ERM	Effect range median
EU	European Union
EU-SCF	European Union, Scientific Committee for Food
FAO	Food and Agricultural Organization of the United Nations
FATE	Fate and Treatability Estimator (model)
FIS	Fluoroimmunosensor
FOP	Fiber optic probe
FS	Final sludge
FWHM	Full width at half maximum
GC	Gas chromatography
GC-MS or GS/MS	Gas chromatography with mass spectroscopy
GC/MS-MS	Gas chromatography with tandem mass spectroscopy
GPC	Gel permeation chromatography
HAP	Hazardous air pollutant
HIV	Human immunodeficiency virus
HMSO	Her Majesty's Stationery Office, UK

HMW	High molecular weight
HOMO	Highest occupied molecular orbital
HPLC	High-performance liquid chromatography
HPLC-DAD	High-performance liquid chromatography with diode-array detection
HPLC-FI	High-performance liquid chromatography with fluorescence detection
HRT	Hydraulic retention time
I&I	Infiltration and inflow
IARC	International Agency for Research on Cancer
ICCD	Intensified charge-coupled device
ISO	International Organization for Standardization
JEFCA	Joint FAO/WHO Expert Committee on Food Additives
JRC-IRMM	Joint Research Centre, Institute for Reference Materials and Measurements (EC)
LETRSS	Laser-excited time-resolved Shpol'skii spectroscopy
LLE	Liquid-liquid extraction
LMW	Low molecular weight
LOD	Limit of detection
LOQ	Limit of quantification
LUMO	Lowest unoccupied molecular orbital
MBR	Membrane bioreactor
MCL	Maximum contaminant level
MCM	Million cubic meters
MDL	Method detection limit
MGD	Million gallons per day
MGP	Manufactured gas plant
MITI	Ministry of International Trade and Industry (Japan)
MLSS	Mixed liquor suspended solids
NF	Nanofiltration
NH_3-N	Ammoniacal nitrogen
NPDES	National Pollutant Discharge Eliminations System (United States)
NWMP	National Waste Minimization Program (USEPA)
OECD	Organization for Economic Cooperation and Development
OLR	Organic loading rate
PAH	Polycyclic aromatic hydrocarbon
PANH	Polycyclic aromatic nitrogen heterocycle
PAS	Photoelectric aerosol sensor
PCB	Polychlorinated biphenyl
PFE	Pressurized fluid extraction
POP	Persistent organic pollutant
ppb	Parts per billion
ppt	Parts per trillion
PrT	Prethickening

PS	Primary sludge
PVC	Polyvinyl chloride
PW	Produced water
PWS	Prince William Sound, Alaska
QCM	Quartz crystal microbalance
RCV	Rapid cyclic voltammetry
RIVM	National Institute for Public Health and Environment (Netherlands)
RO	Reverse osmosis
RTF	Room temperature fluorescence
SAM	Self-assembled monolayer
SCE	Saturated calomel electrode
SCF	Sludge concentration factor
SERS	Surface-enhanced Raman spectroscopy
SFE	Supercritical fluid extraction
SIM	Selective ion method
sOUR	Specific oxygen uptake rate
SPE	Solid-phase extraction
SPNE	Solid-phase nanoextraction
SPR	Surface plasmon resonance
SRT	Solids retention time
SS	Secondary sludge
SWCNT	Single-walled carbon nanotube
TCA	Tricarboxylic acid
TKN	Total Kjeldahl nitrogen
TLCR	Total lifetime carcinogenic risk
TREEM	Time-resolved excitation-emission matrix
TSS	Total suspended solid
USEPA	U.S. Environmental Protection Agency
UV	Ultraviolet
UVF	Ultraviolet fluorescence (spectroscopy)
UV-VIS	Ultraviolet-visible absorption
VOC	Volatile organic carbon
VSC	Volatile sulfide compound
VSS	Volatile suspended solid
WFD	Water Framework Directive (EU)
WHO	World Health Organization
WTM	Wavelength time matrix
WWTP	Wastewater treatment plant

1

Introduction

Amy J. Forsgren

Xylem Inc., Sundbyberg, Sweden

CONTENTS

1.1 What Are PAHs?

Polycyclic aromatic hydrocarbons (PAHs) are a class of organic compounds that are made up of two or more fused aromatic rings. PAHs are created primarily by incomplete combustion of organic matter: the burning of fossil fuels such as coal, oil, and gas, or biomass such as garbage or sewage sludge, or forest fires.

The variety of combustion/pyrolysis processes and the vast number of organic matter that can be burned add up to a plethora of PAH compounds that can be formed and released. The European Food Safety Authority estimates that there are about 500 PAHs that have been detected in ambient air (EFSA 2008).

Is this a new problem? Well, yes and no. PAHs can occur naturally—e.g., during forest fires or by volcanoes—or as a result of human activity. There are a lot of data indicating that human activity is the major source. Vikelsoe et al. (2002) have studied sediment cores in Denmark dating back to 1914. They report low levels of contamination before World War II, after which a significant rise occurs. Studies of sediment cores at Admiralty Bay, Antarctica, have shown that the highest concentrations of PAHs occurred in the last 30 years. This is attributed to (1) increased industrial activity in South America and (2) more research stations in the area (Martins et al. 2010).

1.2 Why Are PAHs a Concern?

PAHs are high-concern pollutants because they are persistent—they stay in the environment for a long period—and because some of them have been identified as carcinogens, mutagens, or teratogens. One PAH, benzo(a)pyrene or B[a]P, has the dubious distinction of being the first chemical identified as a carcinogen (Sternbeck 2011).

Whether people exposed to PAHs will suffer harmful effects, and what those harmful effects will be, depends, of course, on many factors (ATSDR 1995):

The dose and duration of PAH exposure

The pathway by which the person is exposed—breathing, eating, drinking, skin contact

Other chemicals to which the person is exposed

Individual characteristics, such as age, sex, state of health, and nutritional status

There is an extensive literature on the accumulation of PAHs in mussels and fish (EFSA 2008). Conventional wisdom is that PAHs are accumulated in terrestrial and aquatic plants, fish, and invertebrates, but that many animals are able to metabolize and eliminate PAHs. Bioconcentration factors—the concentration in tissues compared to the concentration in media—are very high for fish and crustaceans, often in the 10 to 10,000 range. Studies of PAH bioconcentration in higher organisms are not so plentiful, and indeed, conventional wisdom has not deemed it to be particularly needed.

A Canadian study, however, published in August 2014 has seen mutations in cormorant chicks that are linked to PAHs (King et al. 2014). This is disturbing and calls for more study.

1.3 Which PAHs Are a Concern?

In the literature, it is frequently noted that one of the difficulties in comparing different reports of PAH measurements is that there is a lack of consistency about which PAHs are included in the measurements taken. Table 1.1 illustrates one reason for this lack of consistency: various agencies have different recommendations for which PAHs to monitor.

Another contributing reason may be that the type of biomass being burned, and how it is burned, will dramatically affect the composition of the PAHs created. Table 1.2 gives examples of the varying relative amounts of four PAHs, estimated as ratios of benzo(a)pyrene.

TABLE 1.1

PAHs Frequently Monitored according to Recommendations/Requirements by Various Agencies

Chemical	CAS Number	EPA, Σ16 PAHs	EPA, Priority Chemical List	ATSDR	EU-SCF, 2002	EU-SCF, 2005	JECFA
Acenaphthene	83-32-9	X	X	X			
Acenaphthylene	208-96-8	X	X	X			
Anthracene	120-12-7	X	X	X			
Benz(a)anthracene	56-55-3	X		X	X	X	X
Benzo(g,h,i)perylene	191-24-2	X	X	X	X	X	
Benzo(a)pyrene	50-32-08	X		X	X	X	X
Benzo(e)pyrene	192-97-2			X			
Benzo(b)fluoranthene	205-99-2	X		X	X	X	X
Benzo(j)fluoranthene	205-82-3			X	X	X	X
Benzo(k)fluoranthene	207-08-9	X		X	X	X	X
Benzo(c)fluorene	205-12-9	X				X	X
Chrysene	208-01-9	X		X	X	X	X
Cyclopenta(c,d)pyrene	27208-37-3				X	X	X
Dibenz(a,h)anthracene	53-70-3	X	X	X	X	X	X
Dibenzo(a,e)pyrene	192-65-4				X	X	X
Dibenzo(a,h)pyrene	189-64-0				X	X	X
Dibenzo(a,i)pyrene	189-55-9				X	X	X
Dibenzo(a,l)pyrene	191-30-0				X	X	X
Fluoranthene	206-44-0	X		X			
Fluorene	86-73-7	X		X			
Indeno(1,2,3-c,d)pyrene	193-39-5	X	X	X	X	X	X
5-Methylchrysene	3697-24-3				X	X	X

(Continued)

TABLE 1.1 (Continued)
PAHs Frequently Monitored according to Recommendations/Requirements by Various Agencies

Chemical	CAS Number	EPA, Σ16 PAHs	EPA, Priority Chemical List	ATSDR	EU-SCF, 2002	EU-SCF, 2005	JECFA
Naphthalene	91-20-3	X					
Phenanthrene	85-01-8	X	X	X			
Pyrene	129-00-0	X	X	X			

Source: Compiled from EPA, Polycyclic Aromatic Hydrocarbons, EPA Fact Sheet, U.S. Environmental Protection Agency, Office of Solid Waste, Washington, DC, January 2008; ATSDR, Polycyclic Aromatic Hydrocarbons (PAHs) ToxFAQ, U.S. Department of Health and Human Services, Agency for Toxic Substances and Disease Registry, Atlanta, GA, 1996, http://www.atsdr.cdc.gov/toxfaq.html; JRC-IRMM, Polycyclic Aromatic Hydrocarbons (PAHs) Factsheet: 3rd Edition, JRC 60146-2010, European Commission, Joint Research Centre, Institute for Reference Materials and Measurements, Geel, Belgium, 2010; EFSA, *The EFSA Journal*, 724, 1–114, 2008; EC, European Commission, Opinion of the Scientific Committee on Food, 2002, http://europa.eu.int/comm/food/fs/sc/scf/out153_en.pdf (accessed August 30, 2014); EC, *Official Journal of the European Commission*, L34, 43, 2005; Wenzl et al., *TrAC Trends in Analytical Chemistry*, 25(7), 716–725, 2006.

Note: CAS, Chemical Abstracts Service; EPA, U.S. Environmental Protection Agency; EU-SCF, EU Scientific Committee for Food; ATSDR, U.S. Department of Health and Human Services, Agency for Toxic Substances and Disease Registry; JECFA, Joint FAO/WHO Expert Committee on Food Additives.

TABLE 1.2

PAH Profiles for Various Sources, Estimated as a Ratio to B[a]P

Source	Stationary or Mobile?	Benzo(b) Fluoranthene	Benzo(k) Fluoranthene	Benzo(a) Pyrene	Indeno(1,2,3-cd) Pyrene
Coal combustion (industrial and domestic)	Stationary	0.05	0.01	1.0	0.8
Wood combustion (industrial and domestic)	Stationary	1.2	0.4	1.0	0.1
Natural fires/ agricultural biomass burning	Stationary	0.6	0.3	1.0	0.4
Anode baking	Stationary	2.2[a]	[a]	1.0	0.5
Passenger cars, conventional	Mobile	1.2	0.9	1.0	1.0
Passenger cars, closed-loop catalyst	Mobile	0.9	1.2	1.0	1.4
Passenger cars, diesel, direct injection	Mobile	0.9	1.0	1.0	1.1
Passenger cars, diesel, indirect injection	Mobile	0.9	0.8	1.0	0.9
Heavy-duty vehicles	Mobile	5.6	8.2	1.0	1.4

Source: EFSA, *The EFSA Journal*, 724, 1–114, 2008.

[a] Combined result for the two PAHs.

1.4 Why Wastewater Treatment Plants?

With the exception of sludge incineration, wastewater treatment plants do not create PAHs. PAHs are brought into the wastewater treatment plant (WWTP) in the raw influent stream.

PAHs are created by many mobile and stationary sources; in large urban areas with millions of motor vehicles, the number of sources generating PAHs can easily number in the millions. Through the sewer systems, PAHs are collected and directed into the WWTP influent stream. Some PAHs will be broken down in the WWTP processes, some water-soluble PAHs will exit in the treated effluent stream, and some will adsorb onto particles and be concentrated in the sludge stream.

The WWTP is thus a major point source for collection, concentration, and discharge of PAHs. Control devices, equipment, and methods implemented here for PAH remediation can have a significant environmental impact.

1.4.1 Collection into the WWTP Influent

PAHs are created during combustion; for the most part, they enter the atmosphere in the gas phase, become adsorbed onto particles, and eventually are deposited on land or in water. PAHs generated by automobiles deposit very quickly, on the road or close to it. Rain then washes them, as road run-off or street dust, into the storm sewers, where they make their way to the WWTP. For urban watersheds, this is a major source: Takada et al. (1991) measured Tokyo street dust and found that it can contain PAHs on the order of a few μg/g. In Los Angeles, California, the average traffic is greater than 81 million vehicle-miles per day. This translates to a yearly estimate of 740 kg of PAHs discharged to the waters of the southern California Bight (Stein et al. 2006)—and that is not counting the PAHs that end up in the sewage sludge.

Research has shown repeatedly that sewage sludge is a very efficient sorbent for all lipophilic contaminants that find their way into the sewage system (Strandberg et al. 2001). PAHs have a lipophilic nature, and thus are concentrated strongly in the sewage sludge. There is increasing environmental concern over fates of pollutants in the solid wastes generated by wastewater treatment processes.

1.4.2 PAHs Generated by WWTPs

PAHs are generated during incineration of organic matter, including sewage sludge incineration (Mininni et al. 2004; Sun 2011). For more discussion on this, please see Chapter 10.

Other mechanisms by which WWTPs can generate PAHs have been proposed, but the amounts seem to be insignificant compared to those generated by incineration.

Off-gases from aeration basins are potentially another source of PAHs generated by WWTPs. Two mechanisms can be expected: volatile species, including low molecular weight PAHs, partitioning into the aeration gas, or particulate matter (with adsorbed PAHs) being thrown into the atmosphere by bubbles bursting at the liquid surface. Upadhyay et al. (2013) have measured particulate matter emissions from aeration basins, with and without odor control, at WWTPs in Arizona. They demonstrated that aerosolization of wastewater occurs, but that aeration basins are not a significant source of particulate matter mass or PAHs associated with particulates (though they found that the finer particles travel beyond the WWTP boundaries, with possible implications for carrying disease-causing agents). Manoli and Samara (2008) also estimate that the amount of PAHs released into the atmosphere via this route is small; only an estimated 1 to 2% of

PAHs are removed by volatization in conventional wastewater treatment plants. Kappen (2003) has also looked at air samples of the off-gas from aeration tanks. She reports that volatilization of the PAHs present was minimal and "should not be a concern unless there are unusually high concentrations of PAHs in the influent."

There are also some indications that the prevailing aerobic and sunlight conditions of sedimentation ponds can transform PAHs into oxygenated PAHs or oxy-PAHs (Kalmykova et al. 2014). The oxy-PAHs are of interest because they are toxic to both humans and the environment, persistent, and more water soluble (and therefore more mobile) than their corresponding PAHs (Lundstedt et al. 2007). This is an area that deserves more attention.

1.5 Why This Matters

A group of scientists in Ontario Province, Canada, studied double-crested cormorants in three Canadian colonies, two in Hamilton Harbour and the third in the cleaner northeastern Lake Erie. Hamilton Harbour is one of the most polluted sites on the Great Lakes, with very high concentrations of PAHs in sediments and the air. During the 2 years of their study, industrial PAH emissions in the area were in the range of thousands of kilograms per year. Levels of mutations in chicks were up to sixfold higher in Hamilton Harbour; bile and liver analysis revealed the PAH benzo(a)pyrene. The inference is that the cormorants are exposed to PAHs and metabolizing them, and the PAHs in turn are causing the observed mutations (King et al. 2014).

Their report, published in August 2014, may be the first one documenting PAH metabolites in wild birds that are caused by ambient chemical contamination, rather than oil spills.

Unfortunately, it may be the first of many.

References

ATSDR. (1995). Toxicological profile for polycyclic aromatic hydrocarbons. U.S. Department of Health and Human Services, Agency for Toxic Substances and Disease Registry, Atlanta, GA.

ATSDR. (1996). Polycyclic aromatic hydrocarbons (PAHs) ToxFAQ. U.S. Department of Health and Human Services, Agency for Toxic Substances and Disease Registry, Atlanta, GA. http://www.atsdr.cdc.gov/toxfaq.html.

EC. (2002). European Commission, Opinion of the Scientific Committee on Food, 2002. http://europa.eu.int/comm/food/fs/sc/scf/out153_en.pdf (accessed August 30, 2014).

EC. (2005). European Union, Commission Recommendation 2005/108/EC. *Official Journal of the European Commission*, L34, 43.

EFSA. (2008). Polycyclic aromatic hydrocarbons in food: Scientific opinion of the Panel on Contaminants in the Food Chain (question no. EFSA-Q-2007-136). *The EFSA Journal*, 724, 1–114.

EPA. (2008, January). Polycyclic aromatic hydrocarbons. EPA Fact Sheet. U.S. Environmental Protection Agency, Office of Solid Waste, Washington, DC.

JRC-IRMM. (2010). Polycyclic aromatic hydrocarbons (PAHs) factsheet: 3rd edition. JRC 60146-2010. European Commission, Joint Research Centre, Institute for Reference Materials and Measurements, Geel, Belgium.

Kalmykova, Y., Moona, N., Strömvall, A.M., and Björklund, K. (2014). Sorption and degradation of petroleum hydrocarbons, polycyclic aromatic hydrocarbons, alkylphenols, bisphenol A and phthalates in landfill leachate using sand, activated carbon and peat filters. *Water Research*, 56, 246–257.

Kappen, L.L. (2003). Volatilization and fate of polycyclic aromatic hydrocarbons during wastewater treatment. Thesis, College of Engineering, University of Cincinnati, Cincinnati, OH.

King, L.E., de Solla, S.R., Small, J.M., Sverko, E., and Quinn, J. (2014). Microsatellite DNA mutations in double-crested cormorants (*Phalacrocorax auritus*) associated with exposure to PAH-containing industrial air pollution. *Environmental Science and Technology*. DOI: 10.1021/es502720a.

Lundstedt, S., White, P.A., Lemieux, C.L., Lynes, K.D., Lambert, I.B., Öberg, L., Haglund, P., and Tysklind, M. (2007). Sources, fate, and toxic hazards of oxygenated polycyclic aromatic hydrocarbons (PAHs) at PAH-contaminated sites. *AMBIO: A Journal of the Human Environment*, 36(6), 475–485.

Manoli, E., and Samara, C. (2008). The removal of polycyclic aromatic hydrocarbons in the wastewater treatment process: Experimental calculations and model predictions. *Environmental Pollution*, 151(3), 477–485.

Martins, C.C., Bícego, M.C., Rose, N.L., Taniguchi, S., Lourenço, R.A., Figueira, R.C., Mahiques, M.M., and Montone, R.C. (2010). Historical record of polycyclic aromatic hydrocarbons (PAHs) and spheroidal carbonaceous particles (SCPs) in marine sediment cores from Admiralty Bay, King George Island, Antarctica. *Environmental Pollution*, 158(1), 192–200.

Mininni, G., Sbrilli, A., Guerriero, E., and Rotatori, M. (2004). Polycyclic aromatic hydrocarbons formation in sludge incineration by fluidised bed and rotary kiln furnace. *Water, Air and Soil Pollution*, 154(1–4), 3.

Stein, E.D., Tiefenthaler, L.L., and Schiff, K. (2006). Watershed-based sources of polycyclic aromatic hydrocarbons in urban storm water. *Environmental Toxicology and Chemistry*, 25(2), 373–385.

Sternbeck, J. (2011). *Using sludge on arable land: Effect based levels and longterm accumulation for certain organic pollutants*. Nordic Council of Ministers.

Strandberg, L., Johansson, M., Palmquist, H., and Tysklind, M. (2001). The flow of chemicals within the sewage systems—Possibilities and limitations for risk assessment of chemicals. *URBAN Water*, Umeå University, Umeå, Sweden.

Sun, K. (2011). Examination of the human health issues of sewage sludge and municipal solid waste incineration in North Carolina. Blue Ridge Environmental Defense League (USA). http://bredl.org (accessed August 17, 2014).

Takada, H., Onda, T., Harada, M., and Ogura, N. (1991). Distribution and sources of polycyclic aromatic hydrocarbons (PAHs) in street dust from the Tokyo metropolitan area. *Science of the Total Environment*, 107, 45–69.

Upadhyay, N., Sun, Q., Allen, J.O., Westerhoff, P., and Herckes, P. (2013). Characterization of aerosol emissions from wastewater aeration basins. *Journal of the Air and Waste Management Association*, 63(1), 20–26.

Vikelsoe, J., Thomsen, N., Carlson, L., and Johansen, E. (2002). Persistent organic pollutants in soil, sludge and sediment: A multianalytical field study of selected organic chlorinated and brominated compounds. NERI Technical Report 402. National Environment Research Institute, Aarhus, Denmark.

Wenzl, T., Simon, R., Anklam, E., and Kleiner, J. (2006). Analytical methods for polycyclic aromatic hydrocarbons (PAHs) in food and the environment needed for new food legislation in the European Union. *TrAC Trends in Analytical Chemistry*, 25(7), 716–725.

2

PAHs in Natural Waters: Natural and Anthropogenic Sources, and Environmental Behavior

Jan Kochany

Mississauga, Ontario, Canada

CONTENTS

Polycyclic aromatic hydrocarbons (PAHs) are a class of toxic pollutants that have accumulated in the environment due to both natural and anthropogenic activities (JRC, 2010). The main sources of PAHs can be classified as either pyrogenic or petrogenic (Sanders et al. 2002; Dahle et al. 2003). Pyrogenic-derived PAHs originate from oxygen-depleted, high-temperature processes such as incomplete combustion, pyrolysis, cracking, and destructive distillation. Petrogenic-derived PAHs originate from petroleum, including crude oil, fuels, lubricants, and derivatives of those materials. The composition of PAHs from pyrogenic and petrogenic sources differs. Pyrogenic PAHs are composed of approximately 60% alkyd compounds, while petrogenic PAHs are composed of up to 99% alkyd congeners (Hawthorne et al. 2006). PAHs enter the aquatic environment from incomplete combustion of organic materials, fossil fuels, petroleum product spillage, and various industrial activities, as well as from natural processes such as forest fires and volcanic eruptions. Human exposure to PAHs can take place through multiple routes, including air, soil, food, water, and occupational exposure. They have detrimental effects on the flora and fauna of affected habitats, resulting in the uptake and accumulation of toxic chemicals in food chains and potential serious health problems or genetic defects in humans (WHO 1998; WHO 2000; USEPA 2001; Srogi 2007).

2.1 Properties of PAHs and Environmental Regulations

2.1.1 Chemical Structures and Properties of PAHs

PAHs molecules consist of two or more fused benzene and pentacyclic rings in linear, angular, or cluster arrangements. Several hundred PAHs have been identified (Fazlurrahman et al. 2008). They have been classified into two groups:

  Low molecular weight (LMW), containing three or fewer aromatic rings

  High molecular weight (HMW), containing four or more aromatic rings

The parent structures of several of the most commonly studied PAHs are shown in Figure 2.1.

The high hydrophobicity and chemical stability of PAHs make them persist in the environment. Their hydrophobicity generally increases with increasing molecular mass, with aqueous solubility declining from the low mg/l range for LMW PAHs to about 1 µg/l for HMW PAHs (Pearlman et al. 1984; Wilson and Jones 1993). In addition to the variable structure of the PAH ring system, the molecules may carry various side chains instead of

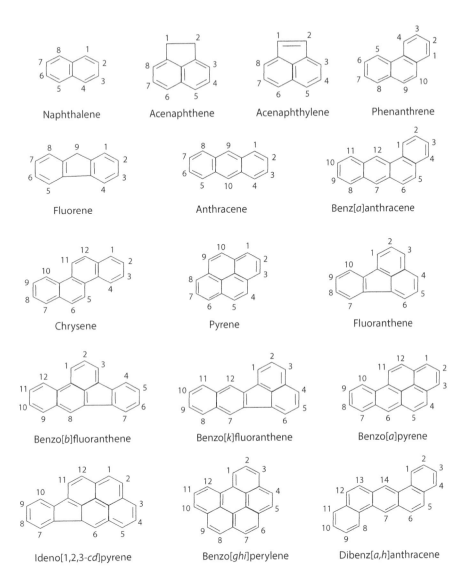

FIGURE 2.1
Parent structures of representative PAHs. (Reprinted with permission from Yan et al., *Mutation Res.* 557(1): 99–108, 2004.)

hydrogen atoms. The properties of representative PAHs are presented in Table 2.1.

Pure PAHs are usually crystalline solids at ambient temperature. PAHs are readily adsorbed to surfaces in aquatic environments or to soil and dust particles, which could be distributed through air (WHO 2000). In the aquatic environment PAH solubility, transport, and fate are dependent on

TABLE 2.1

Properties of 16 PAHs Listed by the USEPA

Compound	CAS Number	Molecular Weight	Log Kow	Water Solubility at 25°C (mg/L)	Melting Point (°C)	Vapor Pressure at 25°C (mPa)
Naphthalene	91-20-3	128	3.5	31.7	80.5	11960
Acenaphthene	83-32-9	154	4.33	3.42	95	594
Fluorene	86-73-7	166	4.18	1.98	116.5	94.7
Phenanthrene	85-01-8	178	4.5	1.29	101	90.7
Anthracene	120-12-7	178	4.5	0.045	216	25
Pyrene	129-001	202	4.9	0.135	156	91.3×10^{-6}
Fluoranthene	206-44-0	202	5.1	0.26	111	328
Chrysene	218-01-9	228	5.3	0.088	158	82.2×10^{-6}
Benz[a]anthracene	56-66-3	228	5.6	0.0057	162	14.7×10^{-3}
Benzo[a]pyrene, BaP	50-32-8	252	6.0	0.0038	179	0.37×10^{-6}
Benzo[b]fluoranthene	205-99-2	252	6.06	0.014	168	0.13×10^{-5}
Benzo[f]fluoranthene	205-82-3	252	6.0	0.008	166	0.18×10^{-5}
Benzo[k]fluoranthene	207-08-9	252	6.06	0.0043	217	2.8×10^{-9}
Indeno[1,2,3-c,d]pyrene	193-39-5	276	6.4	0.00053	164	1.3×10^{-5}
Benzo[g,h,i]perylene	191-24-2	276	5.8	0.003	152	3.5×10^{-5}
Dibenzo[a,h]anthracene	53-70-3	278	5.2	0.0008	176	1.1×10^{-5}

Source: USEPA, *Emergency Planning and Community Right-to-Know Act—Section 313: Guidance for Reporting Toxic Chemicals: Polycyclic Aromatic Compounds Category*, EPA 260-B-01-03, U.S. Environmental Protection Agency, Office of Environmental Information, Washington, DC, August 2001; USEPA, Appendix A to 40 CFR, Part 423—126 Priority Pollutants, U.S. Environmental Protection Agency, Washington, DC, 2009, http://www.epa.gov/NE/npdes/permits/generic/prioritypollutants.pdf (accessed April 14, 2014).

the presence of humic substances that may adsorb and incorporate them into the fulvic and humic acid molecules (Senesi and Miano 1995).

PAHs have characteristic UV absorbance spectra. Each ring structure has a unique UV spectrum; thus, each isomer has a different UV absorbance spectrum. This is especially useful in the identification of PAHs. Most PAHs are also fluorescent, emitting characteristic wavelengths of light when they are excited (when the molecules absorb light). For that reason, UV fluorescence (UVF) spectroscopy is a screening technique that is used in the ecological risk assessment to rapidly estimate total PAH levels in water and sediments (NFEC 2000).

PAHs almost always occur in the environment as complex mixtures. They are analyzed using gas chromatography coupled with mass spectrometry (GC-MS) or high-performance liquid chromatography (HPLC) with ultraviolet and fluorescence detectors (USEPA 1984; Poster et al. 2006). While chemical analyses provide detailed information about concentrations of individual PAHs, they do not indicate the biological impact of the analyzed contaminants. The need for such ecotoxicological information resulted in developing molecular biomarkers, cellular and physiological parameters that signify exposure to or damage incurred by environmental pollutants. There are several commercially available biomarkers that are used for PAH monitoring in the aquatic environment (CEFAS 2000; Pampanin and Sydnes 2013).

2.1.2 PAH Toxicity

The toxic and carcinogenic properties of substances containing PAHs had been reported even before these compounds were discovered. In the 1880s England, high rates of skin cancer for workers in paraffin refinery and coal tar industries were observed. Later, in the 1920s, it was found that organic extracts of soot are carcinogenic. In 1933, the British chemist E. Kenneway isolated the "coal tar carcinogen" benzo[a]pyrene and demonstrated its carcinogenic activity (Waller 1994). Many PAHs have toxic, mutagenic, and carcinogenic properties. Numerous studies have indicated that one-, two-, and three-ring compounds are acutely toxic (Sims and Overcash 1983; Varanasi 1989), and HMW PAHs are considered to be genotoxic (Lijinsky 1991; Mersch-Sundermann et al. 1992). For unsubstituted PAHs, a minimum of four benzene rings is required to exhibit carcinogenic activity (Pickering 1999). Some PAHs are very weak, while others are strongly carcinogenic (e.g., benzo[a]pyrene). Structure-activity relationships become even more complex when substitution of the molecular structure occurs; for example, although benz[a]anthracene is a fairly weak carcinogen, 7,12-dimethylbenz[a]anthracene is a very potent carcinogen. Also, PAH intermediates produced by incomplete degradation (Kazunga and Aitken 2000) or the photooxidized PAHs (McConkey et al. 1997) are in many cases more toxic than the parent compounds. A mixture of PAHs in the environment can also enhance the genotoxic and carcinogenic potential of individual components (Delistraty 1997).

Since PAHs are highly hydrophobic, they are readily absorbed from the gastrointestinal tract of mammals (Cerniglia and Yang 1984) and are rapidly distributed in a wide variety of tissues, but accumulate primarily in body fat. Toxic effects of PAH exposure have been recently reviewed by the U.S. Agency for Toxic Substances and Diseases Registry (ATSDR 2009).

2.1.3 Environmental Regulations for Drinking Water and Discharges into the Surface Waters

Sixteen parent PAHs are designated by the USEPA and the EU as priority pollutants: naphthalene, acenaphthylene, acenaphthene, fluorene, phenanthrene, anthracene, fluoranthene, pyrene, benz[a]anthracene, chrysene, benzo[b] fluoranthene, benzo[k]fluoranthene, benzo[a]pyrene, dibenz[a,h]anthracene, benzo[g,h,i]perylene, and indeno[1,2,3-c,d]pyrene (Lerda 2010). The limits for

TABLE 2.2

PAH Limits for Drinking Water

Agency	Level (mg/L)	Comments
USEPA	0.0001	MCL for benz[a]anthracene
	0.0002	MCL for benzo[a]pyrene, benzo[b]fluoranthene, benzo[k]fluoranthene, chrysene MCL for PAHs
	0.0003	MCL for dibenzo[a,h]anthracene
	0.0004	MCL for indeno[1,2,3-c,d]pyrene
WHO	0.0007	MCL for benzo[a]pyrene
EU	0.00001	MCL for benzo[a]pyrene
	0.0001	Total concentration of: benzo[b]fluoranthene, benzo[k]fluoranthene, benzo[g,h,i]perylene, indeno[1,2,3-c,d]pyrene

Source: USEPA, *National Primary Drinking Water Regulations*, EPA 816 F-09-004, U.S. Environmental Protection Agency, Washington, DC, May 2009; WHO, *Guidelines for Drinking Water Quality*, 3rd ed., World Health Organization, Geneva, 2008; EU Directive 2008/105, Directive 2008/105/EC of the European Parliament and of the Council on Environmental Quality Standards in the Field of Water Policy, EU.OJL 348/84.

PAH concentrations in drinking water have been established only for a few of them. Maximum contaminant levels (MCLs) for PAHs in drinking water are summarized in Table 2.2.

The criteria for PAH levels compatible with aquatic life have been studied and discussed for many years. Most jurisdictions established criteria for benzo[a]pyrene (BaP) as a proven toxic compound and carcinogen. While the MCL for BaP in water is generally consider to be 0.01 µg/L (EU Directive 2008; USEPA 2014), the MCLs for sediment are not unified.

Recently the National Institute for Public Health and the Environment (RIVM) in the Netherlands published a report (RIVM 2012) with specific ecological risk criteria for 16 EPA-listed PAHs. The report derived risk limits for each individual PAH in water and sediment on the basis of studies conducted in various countries over 25 years. These limits are summarized in Table 2.3.

2.2 PAHs as Environmental Pollutants

PAHs in aqueous samples are present in both dissolved and particulate phases, with LWM PAHs found predominantly in the dissolved phase and HMW PAHs in the particulate phase.

PAHs are released mainly into the atmosphere and are subject to short- and long-range transport, depending on their phases and the size of particulates to which they are associated. From the atmosphere, PAHs are removed

TABLE 2.3

Risk Limits for Individual PAHs in the Aquatic Environment

Compound	Ecosystem Limit				
	$MAC_{eco,water}$ (µg/L)	$MPC_{eco,water}$ (µg/L)	$SRC_{eco,water}$ (µg/L)	$MPC_{eco,sedim}$ (mg/kg)	$SRC_{eco,sedim}$ (mg/kg)
Naphthalene	130	2.0	518	0.16	42
Acenaphthylene	33	1.3	72	0.17	9.5
Acenaphthene	3.8	3.8	102	0.97	31
Fluorene	34	1.5	117	0.83	64
Phenanthrene	6.7	1.1	43	0.78	63
Anthracene	0.10	0.10	4.2	0.047	3.2
Pyrene	0.023	0.023	4.2	1.67	136
Fluoranthene	0.12	0.12	12	4.11	96
Chrysene	0.070	0.070	1.6	1.64	38
Benz[a]anthracene	0.10	0.012	3.1	0.35	91
Benzo[k] fluoranthene		0.017	0.93	0.79	44
Benzo[b] fluoranthene		0.017	1.3	0.79	62
Benzo[a]pyrene	0.010	0.010	0.87	0.49	42
Benzo[g,h,i] perylene	0.0082	0.0082	0.16	0.49	10
Dibenzo[a,h] anthracene	0.014	0.0014	0.14	0.18	18
Indeno[1,2,3-c,d] pyrene		0.0027	0.64	0.38	89

Source: RIVM, *Environmental Risk Limits for Polycyclic Aromatic Hydrocarbons (PAHs): For Direct Aquatic, Benthic, and Terrestrial Toxicity*, RIVM Report 607711007/2012, National Institute for Health and Environment (RIVM), Bilthoven, The Netherlands, 2012.

Note: $MAC_{eco,water}$, maximum acceptable concentration for aquatic ecosystem; $MPC_{eco,water}$, maximum permissible concentration for ecosystem; $SRC_{eco,water}$, serious risk concentration for ecosystem; $MPC_{eco,sedim}$, maximum permissible concentration for water sediment; $SRC_{eco,sedim}$, serious risk concentration for sediment.

by dry and wet deposition into water, soil, and vegetation, where they can undergo volatilization, photolysis, oxidation, biodegradation, adsorption onto particles or sediments, or accumulation in aquatic organisms (Baek et al. 1991). PAHs may be released into the environment through natural phenomena such as forest fires, volcanic eruptions, diagenesis, and biosynthesis (Wilcke et al. 2000), but human activities are considered to be a major source of release of PAHs to the environment (NRC 1983; IPCS 1998; GFEA 2012).

PAHs are widely distributed in the environment and have been detected in numerous media to which humans and biota are exposed, including air, water, food, soil, sediment, and tobacco smoke. PAHs have been detected in rainwater in many countries. Reported concentrations in precipitation are summarized in Table 2.4.

TABLE 2.4

Concentration of PAHs in Rainwater in Various Countries (ng/L)

Country	Oregon, United States	Singapore	Hungary	Germany	Poland	Greece	Italy
Reference	Pankow et al. (1984)	Rianawati (2007)	Kiss et al. (1996)	Levsen et al. (1991)	Grynkiewicz et al. (2002)	Manoli et al. (2000)	Olivella (2006)
Napthalene	8–77	54–806	540–1000	nr	36–269	32–780	nr
Acenaphthene	3.2	25–280	nr	nr	2.9–8.2	nr	6.3
Fluorene	5	180–228	nr	nr	1–7	nr	6.5
Phenanthrene	158–238	190–215	15–23	12–36	1.9–20	29–590	10–200
Anthracene	8–19	39–80	3.2–6	0.7–182	36–148	1.7–96	130–600
Pyrene	77–175	34–120	52–180	11–304	36–148	2–76	25–60
Fluoranthene	115–162	36–94	36–94	23–392	21–74	14–54	67–240
Benz[a]anthracene	20–65	22–34	15–25	3–86	16–51	0.9–12	1.3–4.1
Benzo[a]pyrene	5–36	70–116	10–25	5–187	6.5–15	0.7–11	0.4–2.4
Benzo[b]fluoranthene	5	130–140	2.5–6	3–166	6.5–15	0.3–5.4	1.3–4.8
Benzo[f]fluoranthene	4		6–14	5–17	3–9	2.5–11	0.2–5
Benzo[k]fluoranthene	22	24–30	nr	4–12	3–32	1.4–4	0.6–0.8
Indenol[1,2,3-c,d]pyrene	12	125–130	25–98	6–137	0.8–10	2.1–14	5.2
16 PAHs (EPA)	nr	1800–2400	nr	80–650	nr	nr	nr

Note: nr, not reported.

The World Health Organization (WHO) (2000) considers the concentration of individual PAHs in surface and coastal waters in the neighborhood of 0.05 µg/L to be an important threshold. Concentrations above this point indicate some contamination. The benzo[a]pyrene concentration of 0.7 µg/L corresponds to an excess lifetime cancer risk.

PAHs have been found in surface water all over the world. Concentrations reported from various countries are presented in Table 2.5.

PAHs generally do not absorb light of wavelengths critical to global warming (near-infrared range from 700 to 1000 nm; thermal infrared, between 5 and 20 microns). Unlike substances associated with depletion of stratospheric ozone, they are nonhalogenated compounds of low to moderate persistence in the atmosphere. Given these properties and the low steady-state concentrations of PAHs in the atmosphere, they are not considered to contribute significantly to stratospheric ozone depletion, global warming, or ground-level ozone formation.

TABLE 2.5

Concentration of PAHs in Surface Water of Various Countries (ng/L)

Country	Hungary (mean)	Mexico	Greece (mean)	Iran	Sweden	China (industrial)
Reference	Nagy et al. (2012)	Jaward et al. (2012)	Manoli et al. (2000)	Kafilzadeh et al. (2011)	Karlsson and Viklander (2008)	Zhoua and Maskaouib (2003)
Napthalene	21.09	3.5–12	360	7–49	57–180	47–180
Acenaphthene	3.62	nr	3.6	6–40	1.5–2.0	58–1100
Fluorene	6.16	1.7–18	14	3–18	8.1–11	2–6000
Phenanthrene	29.8	4.6–68	33	6–31	7.6–16	157–1440
Anthracene	2.8	0.5–5.8	1.8	4–32	2.5–1.3	58–1900
Pyrene	14.8	1.8–26.2	9.5	0.8–4.1	8.4–11	43–1430
Fluoranthene	17.8	1.1–14.8	51	3.1–18.6	12–18	34–2700
Benz[a]anthracene	1.9	0.7–9.1	9.1	1.3–7.5	0.2–0.9	115–2456
Benzo[a]pyrene	1.9	0.7–1.8	0.9	1.5–7.2	0.2–0.4	13–4700
Benzo[b]fluoranthene	3.9	0.5–6.4	1.4	1.0–6.2	0.5–0.9	19–680
Benzo[f]fluoranthene	3.2	nr	0.4	1.2–2.6	0.1–0.3	18–920
Benzo[k]fluoranthene	3.6	nr	0.3	0.2–1.1	0.7–0.9	12–610
Indeno[1,2,3-c,d]pyrene	1.6	1.9–5.2	2.7	0.3–0.9	0.1–0.5	73–1105
Dibenzo[a,h]anthracene	1.8	nr	1.3	0.6–6.8	0.1–0.2	66–520
16 PAHs (EPA)	98.5–108	31–133	290–620	51.4–391	118–270	420–10,900

2.3 Natural Sources of PAHs

Forest fires, prairie fires, and agricultural burning contribute the largest volumes of PAHs from natural sources to the atmosphere. The actual amount of PAHs and particulates emitted from these sources varies with the type of organic material burned, type of fire (heading fire vs. backing fire), nature of the blaze (wild vs. prescribed, flaming vs. smoldering), and intensity of the fire. PAHs from fires tend to sorb to suspended particulates and eventually enter the terrestrial and aquatic environments as atmospheric fallout (Eisler 1987). It has been estimated that forest fires represented the single largest source of PAHs to the Canadian environment, releasing about 2010 Mg of PAHs into the atmosphere annually (LGL 1993).

In the atmosphere, PAHs may undergo photolytic and chemical transformations. During this atmospheric entrainment, winds may distribute these particle-sorbed PAHs in a global manner such that they appear even in remote areas of the Arctic or Antarctica (Martins et al. 2010).

PAHs occur naturally in bituminous fossil fuels, such as coal and crude oil (NRC 1983).

The PAH makeup of crude oil and refined petroleum products is highly complex and variable, and no two sources have the same composition (Table 2.6).

Under natural conditions, fossil fuels contribute a relatively small volume of PAHs to the environment because most oil deposits are trapped deep beneath layers of rock. There are, however, petroleum bodies (e.g., tar sands) that, being near the surface, are capable of contributing PAHs to both atmospheric and aquatic environments (Timoney and Lee 2011). In some areas, where the local sources of bitumen or asphalt are located close to the surface, the toxic impact of PAHs associated with these deposits on local environment may be high. For example, 40 PAHs have been found in Channel Islands in South California and in a famous La Brea Tar Pit in Los Angeles (Hostettler et al. 2004; Wärmländer et al. 2011).

There is still some uncertainty as to whether or not biosynthesis of PAH in vegetation, fungi, and bacteria is actually occurring, or whether PAH levels in these organisms have been acquired from other sources. It has recently been found that naphthalene, phenanthrene, and perylene are produced biologically. Biological production of naphthalene has been concluded from its presence in *Magnolia* flowers (Azuma et al. 1996) or flower scents of different Annonaceae species from the Amazon rain forest (Jürgens et al. 2000). *Muscodor vitigenus*, an endophytic fungus growing in the Peruvian Amazon region, has shown production of naphthalene (Daisy et al. 2002). High naphthalene concentrations in *Coptotermes formosanus* termite nests of subtropical North America and nests of various termite genera from tropical Brazil suggest naphthalene synthesis by

TABLE 2.6

Maximum, Minimum, and Mean Content of PAH in
50 Different Crude Oils

Compound	Concentration (mg/kg oil)		
	Minimum	Maximum	Mean
Naphthalene	1.2	3700	875
Acenaphthene	0	58	14
Acenaphthylene	0	11	6
Fluorene	1.4	380	96
Anthracene	0	17	4.6
Phenanthrene	0	400	176
Fluoranthene	1.3	15	6
Pyrene	0	20	9.8
Benz[a]anthracene	0.5	16	6.6
Chrysene	4	120	29
Benzo[b]fluoranthene	0.3	14	4.8
Benzo[k]fluoranthene	0	1.3	0.7
Benzo[a]pyrene	0.2	7.8	1.5
Dibenzo[a,]anthracene	0	7.7	1.25
Benzo[g,h,j]perynene	0	1.8	0.9
Indeno[1,2,3-c,d]pyrene	0	1.8	0.8

Source: Compilation of data from Kerr et al., Polyaromatic
Hydrocarbon Content in Crude Oils around the
World, presented at SPE/EPA International
Petroleum Environmental Conference, Austin, TX,
February 28–March 3, 1999; Wang et al., *Geochem.
Trans.* 15(2), 2014, DOI: 10.1186/1467-4866-15-2.

termites or associated microorganisms (Chen et al. 1998; Wilcke et al. 2000, 2002, 2004). Perylene is known to be produced biologically in anaerobic environments in soils and sediments (Wilcke et al. 2002).

2.4 Anthropogenic Sources of PAHs in the Aquatic Environment

In general, anthropogenic sources of PAHs in the aquatic environment can be divided into two categories: sources that discharge directly into the water body and sources that discharge into the atmosphere. Atmospheric deposition is considered the main source of PAHs in the aquatic environment; therefore, atmospheric emission sources are related to PAH water pollution.

Common sources of PAHs include direct or indirect discharges from petroleum terminals, shipyards, aluminum smelting, manufactured gas production plants, tar distillation plants, rail yards, loading/unloading facilities, and spilled or seeped petroleum or coal- or oil-derived tars and associated distillation products (Battelle 2003).

The load of PAHs directly discharged from industry into the water body is relatively small. In Germany in 2010, the total PAHs emitted by industry was 4107 kg, but only 52 kg (~1.2%) was discharged directly into aquatic environments, mostly with the wastewater from the petrochemical and steel manufacturing industries (GFEA 2012). It is believed, however, that most of the PAHs emitted to the air would end up in the water bodies.

2.4.1 Fossil Fuel Extraction, Refining, and Burning

The total amount of PAHs in coal varied from 1.2 to 28.3 mg/kg for various coal types in the United States (Zhao et al. 2000), from 4 to 36 mg/kg in Poland (Bojakowska and Sokołowska 2001), and from 0.4 to 4.2 mg/kg in Nigeria (Ogala and Iwegbue 2011). PAHs may be leached from coal during its storage and transportation. Greater amounts of PAHs are produced from coal during burning (Liu et al. 2008a). It has been estimated that in 2004 alone, 530,000 tons of the 16 EPA PAHs were emitted into the atmosphere worldwide. China has the lead with 114,000 tons, followed by India with 90,000 tons and the United States with 32,000 tons (Zhang and Tao 2009).

Oil and gas offshore extraction produces large volumes of water, so-called produced water (PW) containing petroleum hydrocarbons as well as PAHs associated with them. Since many oil and gas offshore operations have been developed recently, there is a growing concern regarding their potential for causing adverse effects in the marine environment. The average concentrations of PAHs in PW from North Sea installations are shown in Table 2.7.

Ship-related operational discharges of oil include the discharge of bilge water from machinery spaces, fuel oil sludge, and oily ballast water from fuel tanks. Before international regulations were introduced to prevent oil pollution from ships (MARPOL 73/78), the normal practice for oil tankers was to wash out the cargo tanks with water and then pump the resulting mixture of oil and water into the sea. Also, oil cargo or fuel tanks were used for ballast water, and consequently, oil was discharged into the sea when tankers flushed out the oil-contaminated ballast water. This practice resulted in a heavy charge of PAH contamination on the marine environment.

In addition to the PAH discharges related to normally functioning oil extraction and shipping, substantial environmental damages have occurred during accidents of oil tankers and oil platforms. On March 24, 1989, the T/V *Exxon Valdez* grounded on Bligh Reef in Prince William Sound (PWS), Alaska, discharging about 41 million L of crude oil and polluting 500 km of the shorelines. PAH residue at this site has been reported for many years

TABLE 2.7

Average Concentrations of PAHs in Produced Water from Offshore Operations

Compound	Oil Installation	Gas Installation	Unspecified Installations
Naphtanelene	145	115	108
Phenanthrene	13.6	20.9	10.7
Fluorene	8.3	13.1	6.7
Acenaphthene	2	50.2	1.78
Acenaphthylene	0.86	12.6	2.35
Fluoranthene	0.26	35.4	0.29
Anthracene	3.74	110	1.17
Pyrene	0.63	8	0.47
Benzo[a]pyrene	0.52	—	0.022
Chryzene	0.84	1	0.52
Benz[a]anthracene	0.23	1	0.25
Benzo[b]fluoranthene	0.028	—	0.031
Benzo[k]fluoranthene	0.007	—	0.007
Dibenzo[g,h]anthracene	0.005	—	0.005
Benzo[g,h,j]perylene	0.029	—	0.019
Indeno[1,2,3-c,d]pyrene	0.005	—	0.006

Source: Pampanin and Sydnes, in *Hydrocarbon*, ed. Kutcherov, 2013, DOI: 10.5772/48176.

(Hostettler et al. 1999; Neff 2002). Also, elevated levels of PAHs in water along the shores of Florida, Mississippi, and Louisiana have been found after the April 2010 *Horizon* platform accident in the Gulf of Mexico released 779 million L of crude oil (Allan et al. 2012).

2.4.2 Industrial Sources

Sources of PAHs contributing to industrial emissions include aluminum and coke production, petrochemical industries, rubber tire and cement manufacturing, bitumen and asphalt industries, wood preservation, commercial heat and power generation, and waste incineration.

In 1998 the USEPA published a report about emission factors from various industrial sources in the United States. The data from this report are summarized in Table 2.8.

Studies on the thermal degradation of organic materials revealed that emission factors (EFs) of PAHs ranged from 0.13 to 0.4 mg/g for cellulose and 0.5 to 9.0 mg/g for tire (Fabbri and Vassura 2006; Chen et al. 2007). The reported emissions of PAHs from various industrial stacks demonstrated that EFs of PAHs from these industrial stacks ranged from 0.08 to 3.97 mg/kg feedstock, while EFs for BaP ranged from 1.87 to 15.5 µg/g feedstock. The highest EFs of

TABLE 2.8

Average of Emission Factors (mg PAH per unit of fuel or product) for PAHs in U.S. Industries

PAHs	Emission Factor (mg/Mg)						
	Biosolid Incineration	Hazardous Wastes	Aluminum Industry	Steel Industry	Petrochemistry	Iron Foundries	Coal Combustion
Napthalene	1.59	—	188	0.084	4.3	0.04	—
Acenaphthene	0.23	3.6	—	0.35	4.8	0.06	—
Fluorene	4.40	6.70	63	0.39		0.04	5.0
Phenanthrene	44.0	50.20	187	1.74	94.1	1.7	13.8
Anthracene	0.08	5.8	241	0.19	10.8	0.17	7.0
Pyrene	1.8	14.0	3.55	7.18	68	7.18	5.0
Fluoranthene	62	24.8	2.35	9.82	48.5	9.8	—
Benz[a]anthracene	0.62	3.0	26	3.85	0.79	3.8	4.0
Benzo[a]pyrene	0.51	1.0	7.7	1.29	0.72	0.13	3.0
Benzo[b]fluoranthene	0.07	2.5	4.74	1.40	0.98	1.4	2.0
Benzo[a]fluorene	0.88	—	4.74	1.18		—	—
Benzo[k]fluoranthene	0.61	—	2.3	1.18		—	3.0
Indeno[1,2,3-c,d]pyrene	-0.10	—	4.74	1.58		—	2.0
Dibenzo[a,h]anthracene		0.6	0.474	0.24		—	3.0
Benzo[g,h,i]perylene		0.6	4.74	0.24		—	3.0

Source: Compiled from USEPA, *Locating and Estimating Air Emissions from Sources of Polycyclic Organic Matter*, EPA-454/R-98-014, U.S. Environmental Protection Agency, Office of Air Quality Environmental Protection Planning and Standards, Research Triangle Park, NC, July 1998.

total PAHs and BaP were found from the combustion of heavy oils and coke manufacturing (Yang et al. 1998).

It has been estimated that the global total annual atmospheric emission of 16 PAHs from major sources in 2007 was 504 Gg (Shen et al. 2013), with major contributions as follows: residential/commercial biomass burning, 60.5%; open-field biomass burning (agricultural waste burning, deforestation, and wildfire), 13.6%; and petroleum consumption by on-road motor vehicles, 12.8%. South Asia (87 Gg), East Asia (111 Gg), and Southeast Asia (52 Gg) were the regions with the highest PAH emission densities, contributing half of the global total PAH emissions. Among the global total PAH emissions, 6.19% of the emissions were in the form of high molecular weight carcinogenic compounds, and the percentage of the carcinogenic PAHs was higher in developing countries (6.22%) than in developed countries (5.73%), due to the differences in energy structures and the disparities of technology (Shen et al. 2013).

Recently the European Environmental Agency (EEA) reported (2012) that in 2010, the total PAH emission from EU countries was 1447 Mg, which includes 189 Mg of BaP. The main sources of emission were:

Commercial institutions and households, 59%

Agriculture, 16%

Industrial processes, 9%

Energy use in industry, 5%

2.4.3 Municipal Sources

Mobile sources are major causes of PAH emissions in urban areas. PAHs are emitted from exhaust fumes of vehicles, including automobiles, railways, ships, and aircrafts. PAH emissions from mobile sources are associated with use of diesel, coal, gasoline, oils, and lubricant oil (Baek et al. 1991). It has been reported that the amount of PAHs in engine exhaust decreases with leaner mixtures (Ravindra et al. 2006). Also, total emissions and toxicities of PAHs released from light-duty vehicles using ethanol fuel are less than those using gasohol (Abrantes et al. 2009). It has been reported (He et al. 2010) that diesel engines using diesel/biodiesel and their blends can reduce total emissions of PAHs by 19.4%.

Recent studies (Van Metre and Mahler 2010; Mahler et al. 2012, 2014) revealed that coal tar-based (CT) pavement sealcoat, widely used in the United States, is a source of PAH contamination of water in urban lakes in the United States. CT sealcoat products have also been identified as the main source of PAHs in storm water pond sediment in Minnesota (MPCA 2010) and in the DuPage River in suburban Chicago (Prabhukumar and Pagilla 2010).

The comparison between atmospheric input and output by runoff has shown the importance of street deposits as a source of PAHs for surface waters in urban areas. PAH profiles in the various compartments showed the fate of PAHs in the air-water system: the proportion of carcinogenic PAHs was more important in runoff waters (35%) than in bulk atmospheric deposition (22%) and air (6%) (Motelay-Massei et al. 2006).

2.4.4 Agriculture Sources

Open burning of brushwood, straw, moorland heather, and stubble is the main agricultural source of PAHs. All of these activities involve burning organic materials under suboptimum combustion conditions. Emission factors of PAHs from wood combustion ranged from 16.4 to 1282 mg/kg wood (Schauer et al. 2001). PAH concentrations released from wood combustion depend on wood type, kiln type, and combustion temperature. Eighty to ninety percent of PAHs emitted from biomass burning are LMW PAHs, including naphthalene, acenaphthylene, phenanthrene, fluoranthene, and pyrene. Total emissions of 16 PAHs from the burning of rice and bean straw vary from 9.29 to 23.6 μg/g and from 3.13 to 49.9 μg/g, respectively (Lu et al. 2009).

2.4.5 Contaminated Sites

As a result of unrestricted industrial activity since the Industrial Revolution, and a lack of environmental regulations until the middle of the 20th century, many areas have been contaminated with industrial wastes and are potential sources of water contamination with PAHs. One of the main such contamination sources is former manufactured gas plants (MGPs). MGP sites have been documented to be a source of groundwater contamination for a long time (Mackay and Gschwend 2001).

It has been estimated (Wehrer et al. 2011) that the number of former MGP sites in the United States is 5000, and in the United Kingdom more than 2300; in Germany there are thought to be 1200. Evaluation of PAH contamination at various sites in the United States is presented in Table 2.9.

2.4.6 Extraterrestrial Sources

For over 10 years ground-based and space-based observations have shown infrared emission features characteristic of large (over 100 carbon atoms) PAH molecules in many areas of the universe (Ricca et al. 2012). These observations suggest that PAHs are among the most abundant compounds in the universe, comprising 10 to 20% of all carbon (Allamandola et al. 1989). Analyses of meteorites found in various parts of the Earth also contained PAHs (Plows et al. 2003). Thus, we can expect that some PAHs—probably trace amounts—on Earth have come from outer space.

TABLE 2.9

Concentrations of PAHs in Soil at Contaminated Sites ($\mu g/kg$)

Compound	Wood Preserving	Creosote Production	Wood Treatment	Coking Plant	Gas Works
		Activities			
Acenaphthene	7–1370			29	2–11
Acenaphthlene	5–50	33–77		187	
Anthracene	10–3037	15–693	766	6–130	57–295
Benz[a]anthracene	12–171		356	16–200	155–397
Benzo[a]pyrene	28–82		94		45–159
Benzo[b]fluoranthene	38–140				108–552
Chrysene	38–481	8–1586	321	11–135	183–597
Dibenzo[a,h]anthracene			101	2	950–3836
Fluoranthene	35–1630	21–1464	1350	34	614–3664
Fluorene	3–1792	49–1294	620	7–245	113–233
Indeno[1,2,3-c,d]pyrene	10–23				121–318
Naphthalene	1–3925	5–5769	92	56–59	
Phenanthrene	11–4434	76–3402	1440	27–277	150–716
Pyrene	49–1016	19–1303	963	28–285	170–833

Source: Modified from U.S. Department of Health and Human Services, Toxicological Profile for Polycyclic Aromatic Hydrocarbons, 1995, www.atsdr.cdc.gov/toxprofiles/tp69.pdf.

2.5 Abiotic Environmental Transformations

2.5.1 Chemical Oxidation

Oxidation reactions are viewed as the most effective in PAH degradation. The efficiency of chemical oxidation of PAHs in the aquatic environment depends on these compounds' solubility and on the concentration oxidants, such as singlet oxygen, hydrogen peroxide, and hydroxyl radicals (Kochany and Maguire 1994). These oxidants are known to be produced in water under solar irradiation (Zafiriou et al. 1984; Ross and Crosby 1985). The concentration of oxidizing agents in natural waters will depend on many factors, e.g., solar irradiation, temperature, and content of humic substances (Hoigne et al. 1989). Some inorganic salts and oxides can also be involved in the production of oxidants in natural waters, and thus influence the chemical oxidation of PAHs. The reaction proceeds with complex pathways producing numerous intermediates. However, the final reaction products include a mixture of ketones, quinones, aldehydes, phenols, and carboxylic acids (Rivas et al. 2000; Reisen and Arey 2002).

2.5.2 Photochemical Oxidation

PAHs strongly absorb light in the visible or near-ultraviolet part of the spectrum. Absorption maxima are shifted to longer wavelengths with increasing size of the molecules. Substituents alter the spectral properties of PAHs, but these alterations depend on both the nature of the substituent and its position in the aromatic ring system (Zander 1983). Since most PAHs absorb solar radiation of wavelength greater than 290 nm, they can undergo direct photochemical reactions. Photochemical degradation of PAHs often involves the same oxidative species that are produced during the purely chemical oxidation of PAHs. Consequently, the reaction products include similar complex mixtures (David and Boule 1993; Mallakin et al. 2000).

Photolysis of benzo[a]pyrene and benz[a]anthracene under both simulated and natural solar radiation has been studied by Mill et al. (1981). They found quinones as the main photoproducts.

They have also confirmed the importance of oxygen in water, since when water was purged with nitrogen, the photolysis was strongly inhibited. It has been suggested that PAHs with up to three condensed rings decompose according to the mechanism with a radical cation (fluoranthene, anthracene, naphthalene, acenaphthene, and phenanthrene), and the PAHs with four rings (benz[a]anthracene, benzo[a]pyrene, and chrysene) in radical reactions with the participation of oxygen in the initial reaction period (Miller and Olejnik 2001).

It has been found that toxicity of PAHs to plant and animals increases with near-ultraviolet (280–400 nm) radiation (Arfsten et al. 1996). PAHs and UV light in the presence of oxygen resulted in enhancing the carcinogenic properties of UV light. Carcinogenic response was related to UV dose and PAH concentration. A strong photomutagenic response has been observed for anthracene, benz[a]anthracene, benzo[g,h,i]perylene, benzo[a]pyrene, indeno[1,2,3-c,d]pyrene, and pyrene (Yan et al. 2004). This phenomenon is likely related to structure photomodifications of PAH molecules generating more toxic, carcinogenic, and mutagenic compounds than the parent PAH (Mallakin et al. 2000; Yu 2002).

It has been documented that photooxidation of PAHs is an important pathway of degradation of these compounds in the aquatic environment (Kochany and Maguire 1994). The rate of PAH photodegradation depends on the molecular structure of the compounds, particularly on the number of condensed aromatic rings. Generally, compounds with higher molecular weight and with more condensed aromatic rings photolyze faster.

2.5.3 Biological Transformations

2.5.3.1 Bioaccumulation

Bioaccumulation of PAHs in aquatic organisms, particularly fish, has been studied for many years. Literature data indicate that bioaccumulation of PAH in aquatic organisms occurs at different concentrations, which depend on the molecular weight of the compound, different feeding habits, habitat, and biotransformation capacities of the organisms in relation to trophic levels (Advait et al. 2013). PAH concentrations of up to 44.9 µg/g wet weight have been found in tissues of fish inhabiting aquatic habitats of high anthropogenic influence in the Arabian Gulf. In other cases, levels of PAH in fish samples have ranged between 1.91 and 224.03 ng/g wet weight, recorded in China, and 43 and 195 ng/g wet weight, as recorded in Australia. Particularly high accumulation of PAHs was found in mollusks and crabs: 85 and 80 µg/g, respectively (Ololade and Lajide 2010). Since fish and other marine organisms are an important part of human diet limits for PAHs, concentrations have been established. EC Directive 2006 limits concentration of BaP in mollusks to 10 µg/kg and in fish to 2.0 µg/kg of wet weight.

2.5.3.2 Biodegradation

PAH-degrading microorganisms are widely distributed in the natural environment. A list of microorganisms capable of PAH degradation has been published by Mueller et al. (1996).

Three different aerobic degradation mechanisms of PAHs have been identified (Bamforth and Singleton 2005). They are shown in Figure 2.2. The basis of these mechanisms is the oxidation of the aromatic ring, followed by the systematic breakdown of the compound to PAH metabolites or carbon dioxide. The source of oxygen atoms can be oxygen from the air or water.

Microbial oxygenation by fungi and bacteria involves specialized enzymes (Neilson and Allard 1998):

Oxygenases, catalyzing oxygen incorporation

Hydroxylases, catalyzing oxygen from water incorporation

The complete degradation of such complex structures as PAHs involves a long sequence of reactions. During aerobic biodegradations several intermediate compounds such as catechols, phenols, benzoates, and salicylates are formed. These intermediates are further degraded by ring fission to aliphatic compounds that enter the tricarboxylic acid (TCA) cycle (also known as the citric acid cycle), to provide energy and cell carbon. As can be expected, for higher PAHs with three or more aromatic rings, degradation may be more complex and depends on the following issues:

1. Dioxygenation at different positions of the rings
2. Fission of dihydroxy compounds produced by dehydrogenation
3. Monooxygenation
4. O-methylation of phenols produced from arene oxides

Biodegradation pathways for fluorene and benzo[a]pyrene have been presented in Figures 2.3 and 2.4, respectively.

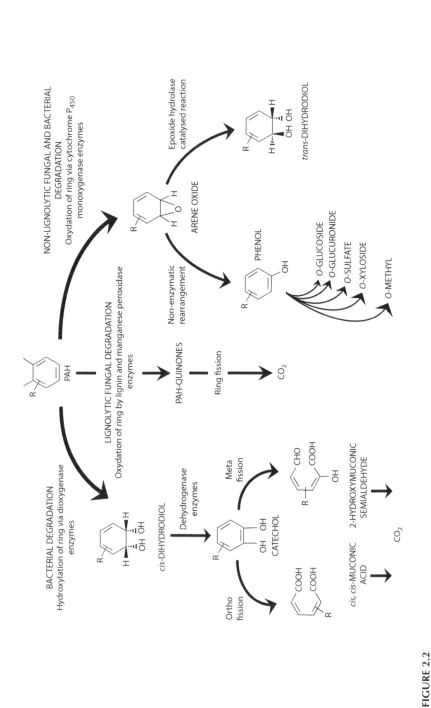

FIGURE 2.2

The main pathway for aerobic PAH degradation by fungi and bacteria. (Reprinted with permission from Cerniglia, *Biodegradation* 3: 351–368, 1992.)

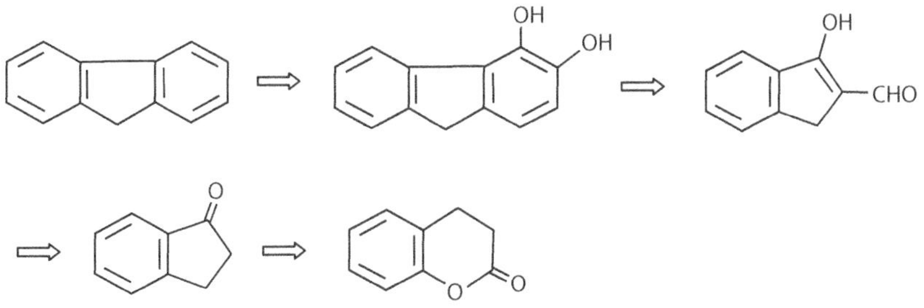

FIGURE 2.3
Aerobic partial biodegradation of fluorene. (Reprinted with permission from Neilson and Allard, in *The Handbook of Environmental Chemistry*, ed. Neilson, Vol. 3, Part J, Springer-Verlag, Berlin, 1998, pp. 1–68.)

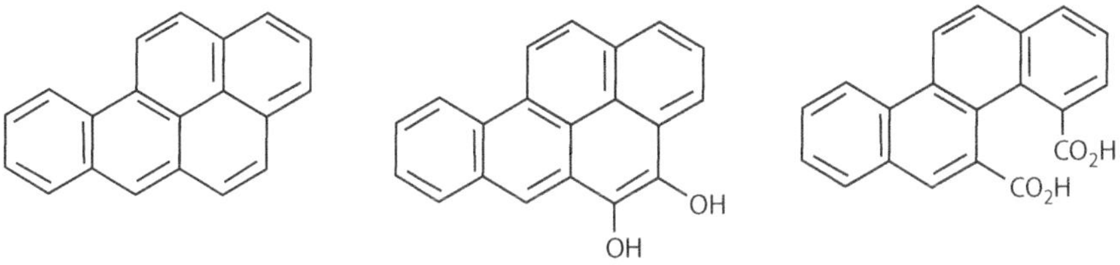

FIGURE 2.4
Aerobic partial degradation of benzo[a]pyrene. (Reprinted with permission from Neilson and Allard, in *The Handbook of Environmental Chemistry*, ed. Neilson, Vol. 3, Part J, Springer-Verlag, Berlin, 1998, pp. 1–68.)

As shown in Figures 2.3 and 2.4, higher PAHs initially undergo only partial biodegradation, and the products of these processes can exist in the aquatic environment for a long time, particularly if they are adsorbed on sediment particles.

It has been found that in the absence of molecular oxygen, alternative electron acceptors such as ferrous iron and sulfate are necessary to oxidize these aromatic compounds. Recent research clearly demonstrated that PAH degradation will occur under both denitrifying (Rockne et al. 2000; Dou et al. 2009; Li et al. 2009) and sulfate-reducing (Meckenstock et al. 2004; Fuchedzhieva et al. 2008; Musat et al. 2009) anaerobic conditions. Several good reviews have been published on biodegradation of PAHs that elaborate on the microbial species capable of degradation, enzymatic mechanisms of these processes (Neilson and Allard 1998; Juhasz and Naidu 2000), and bioremediation of soil and sediment contaminated with PAHs (Mrozik et al. 2003; Bamforth and Singleton 2005; Fazlurrahman et al. 2008; Haritash and Kaushik 2009; Ukiwe et al. 2013).

2.6 Fate of PAHs in Aquatic Environment

2.6.1 Main Processes Affecting PAH Fate in the Aquatic Environment

Because of their low solubility and high affinity for organic carbon, PAHs in aquatic systems are primarily found adsorbed to particles that either have settled to the bottom or are suspended in the water column. It has been estimated that two-thirds of PAHs in aquatic systems are associated with particles, and only about one-third are present in dissolved form (Eisler 1987). PAHs in sediments are mainly bound to natural organic matter, which would be considered to significantly affect the distributions of PAHs in sediments (Zhao et al. 2014). In open waters PAH-laden sediment deposition occurs below the photolytic zone; thus, the conditions for photodecomposition and oxidation are not favorable. Therefore, the primary factor affecting the persistence of deposited PAH is microbial degradation (Cerniglia and Heitkamp 1989). It is generally believed that PAHs sorbed to sediments are only available to organisms after their desorption into the dissolved phase. However, there will always be a dynamic exchange between PAHs in the solid and dissolved phases. It has been reported that the most rapid biodegradation of PAHs occurs at the water-sediment interface (Latimer and Zheng 2003). In general, PAH biodegradation rates in natural sediment and water are inversely related to the number of fused benzene rings in the aromatic nucleus and the number of alkyl groups at the rings. Thus, we can expect that the ratio of LMW/HMW of PAHs contaminating a specific sediment substantially decreases over time, due to faster biodegradation of LMW. The schematics of processes related to PAH fate in aquatic environments are presented in Figure 2.5.

It has been observed that prolonged exposure to PAHs can cause adaptation in microbial populations, resulting in greater resistance to toxicity or enhanced ability to utilize PAHs for metabolism and co-metabolism (Cerniglia and Heitkamp 1989). This phenomenon can be useful for remediation of PAH-laden oil spills in the areas exposed to continuous pollution. It may partly explain why hydrocarbons and PAHs were biodegraded faster in the Gulf of Mexico, which is constantly exposed to the pollution from oil platforms, after the *Horizon* platform accident in 2010, as compared to that after the *Exxon Valdez* accident in Alaska in 1989 (Pampanin and Sydnes 2013).

Generally, volatilization and adsorption to suspended sediments with subsequent deposition are the primary removal processes for medium and high molecular weight PAHs, whereas volatilization and biodegradation are the major removal processes for low molecular weight compounds.

2.6.2 PAH Forensics

Environmental forensic (i.e., fingerprinting) techniques apply ratios of varying kinds of compounds to identify sources. These techniques are used for identifying potentially responsible parties at contaminated sites or evaluating

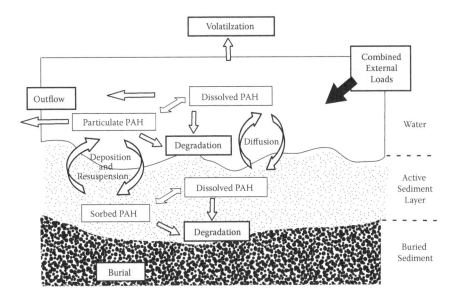

FIGURE 2.5
Fate of PAHs in sediments. (From Greenfield and Davis, A Mass Balance Model for the Fate of PAHs in the San Francisco Estuary, presented at Second Biennial CALFED Science Conference, Sacramento, CA, January 14–16, 2003, http://www.sfei.org/rmp/presentations/GreenfieldPAH.ppt#7.)

the impact of specific human activities on the environment. PAHs are very useful for environmental forensics for the following reasons:

1. They are generated by various natural and anthropogenic sources.
2. They almost always exist in mixtures of compounds whose ratios are characteristic to their source.
3. They are relatively stable in the aquatic environment, particularly in sediments.
4. Components of PAH mixtures degrade in the environment at different rates, which also allows identification of PAH sources.

A list of PAHs that are commonly used to distinguish PAH sources is presented in Table 2.10. The most important factors controlling PAH ratios relate to dry precipitation, surface-to-air diffusion, degradation in air and water, and exchange between water and sediment (Zhang et al. 2005). Zhang et al. (2005) have used a fugacity model to help quantify the differences in transport of individual PAH compounds and to improve the ability to identify sources of PAH contamination using PAH ratios.

PAH ratios work best at sites dominated by a single source. However, most environmental sediment samples have multiple sources of PAHs, for which there may be varying rates of weathering of PAH compounds. Weathering occurs

TABLE 2.10

Ratios of Phenanthrene to Anthracene (P/A) and Fluoranthene to Pyrene (FL/PY) from Different Sources of PAHs

PAH Source	P/A	FL/PY	Reference
Primarily Pyrogenic Sources			
Coke oven emissions	1.27–3.57	0.76–1.31	Maher and Aislabe (1992)
Iron/steel plant (soot)	0.24	0.62	Yang et al. (2002)
Iron/steel plant (flue gas)	0.06	1.43	Yang et al. (2002)
Wood-burning emissions	6.41	1.26	Page et al. (1999)
Auto exhaust soot (gasoline)	1.79	0.90	O'Malley et al. (1996)
Diesel engine soot	0.06	1.26	Bence et al. (1996)
Creosote	0.11–4.01	1.52–1.70	Neff (2002)
Coal tar	3.11	1.29	Neff (2002)
Primarily Petrogenic Sources			
60 crude oils (mean)	52.0	0.25	Kerr et al. (1999)
Australian crude oil	>370[a]	0.78	Neff et al. (2000)
Alaska crude oil	>262[a]	0.2	Bence et al. (1996)
Diesel fuel (no. 2 fuel oil)	>800[a]	0.38	Bence et al. (1996)
Road-paving asphalt	20	<0.11a	Kriech et al. (2002)

Source: Neff et al., *Integr. Environ. Assess. Manage.* 1(1): 22–33, 2005.
[a] Anthracene or fluoranthene concentration was below the detection limit.

primarily through a combination of evaporation/volatilization processes, degradation by microorganisms, and dissolution into water (Battelle 2003). Weathering causes the following key changes to the PAH signature:

Concentrations of LPAHs; (i.e., two- and three-ring) PAHs are reduced, resulting in a decrease in the ratio of LPAHs to four- to six-ring PAHs.

Concentrations of nonalkylated PAHs are reduced, resulting in an increase in the proportion of alkylated PAHs (Battelle 2003).

The use of a variety of source ratios and cross-plots provided the best interpretation of PAH sources in the Fraser River Basin, British Columbia (Yunker et al. 2002). Plots with a random scatter pattern are often indicative of equal weighting of multiple sources of PAHs, whereas less variation in the ratios may indicate that only one or two sources dominate (Yunker et al. 2002).

Mahler et al. (2005) used the PAH source ratios to distinguish coal tar as a PAH source: fluoranthene:pyrene, indeno[1,2,3-c,d]pyrene:benzo[g,h,i]perylene, and benzo[a]pyrene:benzo[e]pyrene. They found that these PAH ratios in suspended sediment collected from four urban streams after storms most closely matched those ratios in particles washed from parking lots with coal tar emulsion sealcoat (Mahler et al. 2005). They also found this set of ratios to be better at distinguishing between the different parking lot samples and stream samples than ratios used for assessing combustion and noncombustion sources. Liu et al. (2008b) used PAH analyses of river sediments to identify petrogenic and pyrogenic sources of pollution.

More sophisticated methodology was applied by Pengchai et al. (2005). They identified the source of PAHs in urban dust using particle size separation into 11 groups based on cluster analysis combined with principal component analysis. They found diesel vehicle exhaust, tire, and pavement were the major contributors of PAHs in the fractionated road dust. Li et al. (2013) identified PAH sources by analyzing the corrected ratios of isomeric PAH species derived using the multimedia fugacity model. Yang et al. (2011) applied an integrated quantitative chemical characterization of PAHs in Alberta oil sand bitumen to distinguish its environmental impact from those resulting from conventional crude oils.

Dated cores of sediments collected from San Francisco Bay were used to reconstruct a history of PAH contamination (Pereira et al. 1999). The ratios of methylphenanthrenes/phenanthrene and (methylfluoranthenes + methylpyrenes)/fluoranthene were sensitive indicators of anthropogenic influences in the estuary. Variations in the ratio of 1,7-dimethylphenanthrene/2,6-dimethylphenanthrene indicated a gradual replacement of wood by fossil fuels as the main combustion source.

2.7 Conclusions

PAHs are important contaminants of the aquatic environment. They are generated by many natural and anthropogenic sources that discharge PAHs directly into the water bodies or indirectly through atmospheric transport and deposition. The most important are anthropogenic sources, particularly related to current and historical fossil fuel extraction, processing, and use.

The ratios of PAHs in their mixture are characteristic to their origin, which allows identification of the source of PAH pollution. In the aquatic environment the majority of PAHs are stored in the sediments and undergo slow degradation, mainly by biological processes.

References

Abrantes, R., Assunção, J.V., Pesquero, C.R., Bruns, R.E., and Nóbrega, R.B. (2009). Emission of polycyclic aromatic hydrocarbons from gasohol and ethanol vehicles. *Atmos. Environ.* 43: 648–654.

Advait, B., Shanta, S., and Abhijit, D.M. (2013). Bioaccumulation kinetics and bioconcentration factors for polycyclic aromatic hydrocarbons in tissues of *Rasbora daniconius*. *IJCPS* 2: 82–94. www.ijcps.org.

Allamandola, L.J., Tielens, A.G.G.M., and Barker, J.R. (1989). Interstellar polycyclic aromatic hydrocarbons: The infrared emission bands, the excitation/emission mechanism, and the astrophysical implications. *Astrophys. J.* 71: 733–775.

Allan, S.E., Smith, B.W., and Anderson, K. (2012). Impact of the Deepwater Horizon oil spill on bioavailable polycyclic aromatic hydrocarbons in Gulf of Mexico coastal waters. *Environ. Sci. Technol.* 46: 2033–2039.

Arfsten, D.P., Schaffer, D.J., and Mulveny, D.C. (1996). The effects of near ultraviolet radiation on the toxic effects of polycyclic aromatic hydrocarbons in animals and plants: A review. *Ecotoxicol. Environ. Safety* 33: 1–24.

ATSDR (Agency for Toxic Substances and Diseases Registry). (2009). Polycyclic aromatic hydrocarbons (PAHs). Atlanta, GA. http://www.atsdr.cdc.gov/csem/csem.html (accessed April 30, 2014).

Azuma, M., Toyota, Y.A., and Kawano, S. (1996). Naphthalene a constituent of *Magnolia* flowers. *Phytochemical* 42: 999–1004.

Baek, S.O., Field, R.A., Goldstone, M.E., Kirk, P.W., and Lester, J.N. (1991). A review of atmospheric polycyclic aromatic hydrocarbons: Sources, fate and behavior. *Water Air Soil Pollut.* 60: 279–300.

Bamforth, S.M., and Singleton, I. (2005). Bioremediation of polycyclic aromatic hydrocarbons: Current knowledge and future directions. *J. Chem. Technol. Biotechnol.* 80: 723–736.

Battelle Memorial Institute. (2003). Guidance for environmental background analysis. Vol. II: Sediment. Appendix A: Chemical fingerprinting of PAHs in sediments recognizing the contribution of urban background. UG-2054-ENV. Prepared by Battelle Memorial Institute, Columbus, OH, Earth Tech, Inc., Honolulu, HI, and NewFields, Inc., Atlanta, GA for Naval Facilities Engineering Service Center, Port Hueneme, CA. http://web.ead.anl.gov/ecorisk/related/documents/Final_BG_Sediment_Guidance.pdf.

Bence, A.E., Kvenvolden, K.A., and Kennicutt, M.C. (1996). Organic geochemistry applied to environmental assessments of Prince William Sound, Alaska, after the *Exxon Valdez* oil spill. *Org. Geochem.* 24: 7–42.

Bojakowska, I., and Sokołowska, G. (2001). Polycyclic hydrocarbons in crude oils from Poland. *Geolog. Q.* 45: 81–86.

CEFAS (Centre for Environment, Fisheries and Aquaculture Science). (2000). Biomarkers of polycyclic aromatic hydrocarbons (PAH) exposure in fish and their application in marine monitoring. Science Series Technical Report Number 110. Lowestoft, UK.

Cerniglia, C.E. (1992). Biodegradation of polycyclic aromatic hydrocarbons. *Biodegradation* 3: 351–368.

Cerniglia, C.E., and Heitkamp, M.A. (1989). Microbial degradation of polycyclic aromatic hydrocarbons (PAH) in the aquatic environment. In *Metabolism of polycyclic aromatic hydrocarbons in the aquatic environment*, ed. U. Varanasi. CRC Press, Boca Raton, FL, pp. 42–68.

Cerniglia, C.E., and Yang, S.K. (1984). Stereoselective metabolism of anthracene and phenanthrene by the fungus *Cunninghamella elegans*. *Appl. Environ. Microb.* 47: 119–124.

Chen, J., Henderson, G., Grimm, C.C., Lloyd, S.W., and Laine, R.A. (1998). Naphthalene in Formosan subterranean termite carton nests. *J. Agric. Food Chem.* 46: 2337–2339.

Chen, S., Su, B., Chang, J.E., Lee, W.J., Huang, K.L., Hsieh, L.T., Huang, J.C., Lin, W.J., and Lin, C.C. (2007). Emissions of polycyclic aromatic hydrocarbons (PAHs) from the pyrolysis of scrap tires. *Atmos. Environ.* 41: 1209–1220.

Dahle, S., Savinov, V.M., Matishov, G.G., Evenset, A., and Naes, K. (2003). Polycyclic aromatic hydrocarbons (PAHs) in bottom sediments of the Kara Sea shelf, Gulf of Ob and Yenisei Bay. *Sci. Total. Environ.* 306: 57–71.

Daisy, B.H., Strobel, G.A., Castillo, U., Ezra, D., Sears, J., Weaver, D.K., and Runyon, J.B. (2002). Naphthalene, an insect repellent, is produced by *Muscodor vitigenus*, a novel endophytic fungus. *Microbiology* 148: 3737–3741.

David, B., and Boule, P. (1993). Phototransformation of hydrophobic pollutants in aqueous medium. 1. PAHs adsorbed on silica. *Chemosphere* 26: 1617–1630.

Delistraty, D. (1997). Toxic equivalency factor approach for risk assessment of polycyclic aromatic hydrocarbons. *Toxicol. Environ. Chem.* 64: 81–108.

Directive 2006/1882/EC. Directive 2006/1882/EC of the European Parliament and of the Council of establishing quality standards for marine products. EU OJL 364 20.12.2006.

Directive 2008/105/EC. Directive 2008/105/EC of the European Parliament and of the Council on environmental quality standards in the field of water policy. EU.OJL 348/84.

Dou, J., Liu, X., and Ding, A. (2009). Anaerobic degradation of naphthalene by the mixed bacteria under nitrate reducing conditions. *J. Hazard. Mater.* 165(1–3): 325–331.

EEA (European Environmental Agency). (2012). *European Union emission inventory report 1990–2010 under the UNECE Convention on Long-Range Trans-Boundary Air Pollution (LRTAP).* Technical Report 8/2012.

Eisler, R. (1987). *Polycyclic aromatic hydrocarbon hazards to fish, wildlife, and invertebrates: A synoptic review.* U.S. Fish and Wildlife Service Biological Report 85(1.11).

Fabbri, D., and Vassura, I. (2006). Evaluating emission levels of polycyclic aromatic hydrocarbons from organic materials by analytical pyrolysis. *J. Anal. Appl. Pyrol.* 75: 150–158.

Fazlurrahman, A.C., Oakeshott, J.G., and Jain, R.K. (2008). Bacterial metabolism of polycyclic aromatic hydrocarbons: Strategies for bioremediation. *Indian J. Microbiol.* 48: 95–113.

Fuchedzhieva, N., Karakashev, D., and Angelidaki, I. (2008). Anaerobic biodegradation of fluoranthene under methanogenic conditions in presence of surface-active compounds. *J. Hazard. Mater.* 153(1–2): 123–127.

GFEA (German Federal Environment Agency). (2012). Polycyclic aromatic hydrocarbons: Harmful to the environment! Toxic! Inevitable? Press Office Wörlitzer Platz 1, D-06844 Dessau-Roßlau, Germany. www.umweltbundesamt.de (accessed April 30, 2014).

Greenfield, B.K., and Davis, J.A. (2003). A mass balance model for the fate of PAHs in the San Francisco estuary. Presented at Second Biennial CALFED Science Conference, Sacramento, CA, January 14–16. http://www.sfei.org/rmp/presentations/GreenfieldPAH.ppt#7.

Grynkiewicz, M., Polkowska, Z.J., and Namiesnik, J. (2002). Determination of polycyclic aromatic hydrocarbons in bulk precipitation and runoff waters in an urban region (Poland). *Atmos. Environ.* 36: 361–369.

Haritash, A.K., and Kaushik, C.P. (2009). Biodegradation aspects of polycyclic aromatic hydrocarbons (PAHs): A review. *J. Hazard. Mater.* 169: 1–5.

Hawthorne, S.B., Miller, D.J., and Kreitinger, J.P. (2006). Measurement of total poly-cyclic aromatic hydrocarbon concentrations in sediments and toxic units used for estimating risk to benthic invertebrates at manufactured gas plant sites. *Environ. Toxicol. Chem.* 25(1): 287–296.

He, C., Ge, Y., Tan, Y., You, K., Han, X., and Wang, J. (2010). Characteristics of poly-cyclic aromatic hydrocarbons emissions of diesel engine fueled with biodiesel and diesel. *Fuel* 89: 2040–2046.

Hoigne, J., Faust, B.C., Haag, W.R., Scully, F.E., and Zepp, R.G. (1989). Aquatic humic substances as sources and sinks of photochemically produced transient reac-tants. In *Aquatic humic substances, influence on the fate and treatment of pollut-ants,* ed. I.H. Suffet and P. MacCarthy. Advances in Chemistry Series 219. ACS, Washington, DC, pp. 363–381.

Hostettler, F.D., Rosenvauer, R.J., and Krenvolden, K.A. (1999). PAH refractory index as a source discriminant of hydrocarbon input from crude oil and coal in Prince William Sound, Alaska. *Org. Geochem.* 30: 873–879.

Hostettler, F.D., Rosenbauer, R.J., Lorenson, T.D., and Dougherty, J. (2004). Geochemical characterization of tarballs on beaches along the California coast. Part I. Shallow seepage impacting the Santa Barbara Channel Islands, Santa Cruz, Santa Rosa, and San Miguel. *Org. Geochem.* 35: 725–746.

IPCS (International Programme on Chemical Safety). (1998). *Environmental health criteria 202: Selected non-heterocyclic polycyclic aromatic hydrocarbons.* WHO, Geneva.

Jaward, F.M., Alegria, H.A., Reyes, J.G.G., and Hoare, A. (2012). Levels of PAHs in the waters, sediments, and shrimps of Estero de Urias, an Estuary in Mexico, and their toxicological effects. *Sci. World J.,* Article ID: 687034. http://dx.doi.org/10.1100/2012/687034 (accessed April 25, 2014).

JRC (Joined Research Centre–European Commission). Technical notes: Polycyclic aromatic hydrocarbons factsheets. JRC 60146-2010.

Juhasz, A.L., and Naidu, R. (2000). Bioremediation of high molecular weight polycyclic aromatic hydrocarbons: A review of the microbial degradation of benzo[a]pyrene. *Int. Biodeter. Biodegr.* 45: 57–88.

Jürgens, A., Webber, A.C., and Gottsberger, G. (2000). Floral scent compounds of Amazonian Annonaceae species pollinated by small beetles and thrips. *Phytochemistry* 55: 551–558.

Kafilzadeh, F., Shiva, H.A., and Malekpour, R. (2011). Determination of polycyclic aromatic hydrocarbons (PAHs) in water and sediments of the Kor River, Iran. *Middle-East J. Sci. Res.* 10(1): 1–7.

Karlsson, K., and Viklander, M. (2008). Polycyclic aromatic hydrocarbons (PAH) in water and sediment from gully pots. *Water Air Soil Pollut.* 188: 271–282.

Kazunga, C., and Aitken, M.D. (2000). Products from the incomplete metabolism of pyrene by polycyclic aromatic hydrocarbon-degrading bacteria. *Appl. Environ. Microb.* 66: 1917–1922.

Kerr, J.M., Melton, H.R., McMillen, S.J., Magaw, R.I., Naughton, G., and Little, G.N. (1999). Polyaromatic hydrocarbon content in crude oils around the world. Presented at SPE/EPA International Petroleum Environmental Conference, Austin, TX, February 28–March 3.

Kiss, G., Varga-Puchony, Z., and Hlavay, J. (1996). Determination of polycyclic aromatic hydrocarbons in precipitation using solid-phase extraction and column liquid chromatography. *J. Chromatogr.* 725: 261–272.

Kochany, J., and Maguire, R.J. (1994). Abiotic transformations of polynuclear aromatic hydrocarbons and polynuclear aromatic nitrogen heterocycles in aquatic environments. *Sci. Total Environ.* 144: 17–31.

Kriech, A.J., Kurek, J.T., Osborn, L.V., Wissel, H.O., Sweeney, B.J. (2002). Determination of polycyclic aromatic hydrocarbons in asphalt and in corresponding leachate water. *Polycycl. Aromat. Comp.* 22: 517–535.

Latimer, J.S., and Zheng, J. (2003). The sources, transport, and fate of PAHs in the marine environment. In *PAHs: An ecotoxicological perspective*, ed. P.E.T. Douben. John Wiley & Sons, Hoboken, NJ, pp. 9–34.

Lerda, D. (2010). Polycyclic aromatic hydrocarbons (PAHs) factsheet. JRC 500871. European Commission, Joint Research Centre, Institute for Reference Materials and Measurements, Geel, Belgium.

Levsen, K., Behnert, S., and Winkeler, H.D. (1991). Organic compounds in precipitation. *Fresenius J. Anal. Chem.* 340: 665–671.

LGL (Lalonde, Girouard, Letendre et Associes). (1993). PAH emissions into the environment in Canada—1990. Prepared for Environment Canada, Conservation and Protection, Quebec Region, Montreal, Quebec.

Li, C.-H., Zhou, H.-W., Wong, Y.-S., and Tam, N.F. (2009). Vertical distribution and anaerobic biodegradation of polycyclic aromatic hydrocarbons in mangrove sediments in Hong Kong, South China. *Sci. Total Environ.* 407(21): 5772–5779.

Li, X., Zhao, T., Zhang, C., Li, P., Li, S., and Zhao, L. (2013). Source apportionment of polycyclic aromatic hydrocarbons in agricultural soils of Yanqing County in Beijing, China. *Environ. Forensics* 14(4): 324–330.

Lijinsky, W. (1991). The formation and occurrence of polynuclear aromatic hydrocarbons associated with food. *Mutat. Res.* 259: 251–262.

Liu, G., Chen, L., Jianfu, Z., Qinghui, H., Zhiliang, Z., and Hongwen, G. (2008b). Distribution and sources of polycyclic aromatic hydrocarbons in surface sediments of rivers and an estuary in Shanghai, China. *Environ. Pollut.* 154: 298–305.

Liu, G., Niu, Z., Van Niekerk, D., Xue, J., and Zheng, L. (2008a). Polycyclic aromatic hydrocarbons (PAHs) from coal combustion: Emissions, analysis, and toxicology. *Rev. Environ. Contam. Toxicol.* 192: 1–28.

Lu, H., Zhu, L., and Zhu, N. (2009). Polycyclic aromatic hydrocarbon emission from straw burning and the influence of combustion parameters. *Atmos. Environ.* 43: 978–983.

Mackay, A.A., and Gschwend, P.M. (2001). Enhanced concentrations of PAHs in groundwater at a coal tar site. *Environ. Sci. Technol.* 35: 1320–1328.

Maher, W.A., and Aislabe, J. (1992). Polycyclic aromatic hydrocarbons in near shore sediments of Australia. *Sci. Tot. Environ.* 112: 143–164.

Mahler, B.J., Van Metre, P.C., Bashara, T.J., Wilson, J.T., and Johns, D.A. (2005). Parking lot sealcoat: An unrecognized source of urban polycyclic aromatic hydrocarbons. *Environ. Sci. Technol.* 39(15): 5560–5566.

Mahler, B.J., Van Metre, P.C., Crane, J.L., Watts, A.W., Scoggins, M., and Williams, E.S. (2012). Coal-tar-based pavement sealcoat and PAHs: Implications for the environment, human health, and stormwater management. *Environ. Sci. Technol.* 46: 3039–3045.

Mahler, B.J., Van Metre, P.C., and Foreman, W.T. (2014). Concentrations of polycyclic aromatic hydrocarbons (PAHs) and azarenes in runoff from coal-tar- and asphalt-sealcoated pavement. *Environ. Pollut.* 188: 81–87.

Mallakin, A., Dixon, D.G., and Greenberg, B.M. (2000). Pathway of anthracene modification under simulated solar radiation. *Chemosphere* 40: 1435–1441.

Manoli, E., Samara, C., Konstantinou, I., and Albanis, T. (2000). Polycyclic aromatic hydrocarbons in the bulk precipitation and surface waters of northern Greece. *Chemosphere* 41: 1845–1855.

MARPOL 73/78. International Convention for the Prevention of Pollution from Shipping. International Marine Organization. www.imo.org (accessed April 30, 2014).

Martins, C.C., Bícego, M.C., Rose, N.L., Taniguchi, S., Lourenço, R.A., Rubens, C.L., Figueira, R.C.L., Mahiques, M.M., and Montone, R.C. (2010). Historical record of polycyclic aromatic hydrocarbons (PAHs) and spheroidal carbonaceous particles (SCPs) in marine sediment cores from Admiralty Bay, King George Island, Antarctica. *Environ. Pollut.* 158: 192–200.

McConkey, B.J., Duxbury, C.L., Dixon, D.G., and Greenberg, B.M. (1997). Toxicity of a PAH photooxidation product to the bacteria *Photobacterium phosphoreum* and the duckweed *Lemna gibba*: Effects of phenanthrene and its primary photoproduct, phenanthrenequinone. *Environ. Toxicol. Chem.* 16: 892–899.

Meckenstock, R.U., Safinowski, M., and Griebler, C. (2004). Anaerobic degradation of polycyclic aromatic hydrocarbons. *FEMS Microbiol. Ecol.* 49(1): 27–36.

Mersch-Sundermann, V., Mochayedi, S., and Kevekordes, S. (1992). Genotoxicity of polycyclic aromatic hydrocarbons in *Escherichia coli* PQ37. *Mutat. Res.* 278: 1–9.

Mill, T., Mabey, W.R., Lan, B.Y., and Baraze, A. (1981). Photolysis of polycyclic aromatic hydrocarbons in water. *Chemosphere* 10: 1281–1290.

Miller, J.S., and Olejnik, D. (2001). Photolysis of polycyclic aromatic hydrocarbons in water. *Water Res.* 35(1): 233–243.

Motelay-Massei, A., Garban, B., Tiphagne-Larcher, K., Chevreuil, M., Ollivon, D. (2006). Mass balance for polycyclic aromatic hydrocarbons in the urban watershed of Le Havre (France): Transport and fate of PAHs from the atmosphere to the outlet. *Water Res.* 40(10): 1995–2006.

MPCA (Minnesota Pollution Control Agency). (2010). Contamination of stormwater pond sediments by polycyclic aromatic hydrocarbons (PAHs) in Minnesota. The role of coal tar-based sealcoat products as a source of PAHs. Saint Paul, MN. www.pca.state.mn.us.

Mrozik, A., Piotrowska-Seget, Z., and Labuzek, S. (2003). Bacterial degradation and bioremediation of polycyclic aromatic hydrocarbons. *Polish J. Environ. Stud.* 12(1): 15–25.

Mueller, J.G., Cerniglia, C.E., and Pritchard, P.H. (1996). Bioremediation of environments contaminated by polycyclic aromatic hydrocarbons. In *Bioremediation: Principles and applications*, ed. R.L. Crawford and D.L. Crawford. Cambridge University Press, Boise, ID, pp. 125–194.

Musat, F., Galushko, A., Jacob, J., Widde, F., Kube, M., Reinhardt, R., Wilkes, H., Schink, B., and Rabus, R. (2009). Anaerobic degradation of naphthalene and 2-methylnaphthalene by strains of marine sulfate-reducing bacteria. *Environ. Macrobiol.* 11: 209–219.

Nagy, A.S., Simon, G., and Vass, I. (2012). Monitoring of polycyclic aromatic hydrocarbons (PAHs) in surface water of Hungarian upper section of Danube River. *Nova Biotechnol. Chim.* 11(1): 27–35.

Neff, J.M. (2002). *Bioaccumulation in marine organisms. Effects of contaminants from oil well produced water*. Elsevier, Amsterdam.

Neff, J.M., Ostazeski, S., Gardiner, W., and Stejskal, I. (2000). Effects of weathering on the toxicity of three offshore Australian crude oils and a diesel fuel to marine animals. *Environ. Toxicol. Chem.* 19: 1809–1821.

Neff, J.M., Stout, S.A., and Gunster, D.G. (2005). Ecological risk assessment of polycyclic aromatic hydrocarbons in sediments: Identifying sources and ecological hazard. *Integr. Environ. Assess. Manage.* 1(1): 22–33.

Neilson, A.H., and Allard, A.-S. (1998). Microbial metabolism of PAHs and heteroarenes. In *The handbook of environmental chemistry*, ed. A.H. Neilson. Vol. 3, Part J, PAHs and Related Compounds. Springer-Verlag, Berlin, pp. 1–68.

NFEC (Naval Facilities Engineering Command). (2000). Near-real time UV fluorescence technique for characterization of PAHs in marine sediment. TechData Sheet. Washington, DC.

NRC (National Research Council). (1983). *Polycyclic aromatic hydrocarbons: Evaluation and effects.* Committee on Pyrene and Selected Analogues, Board on Toxicology and Environmental Health Hazards, Commission on Life Sciences, National Academy Press, Washington, DC.

Ogala, J.E., and Iwegbue, C.M.A. (2011). Occurrence and profile of polycyclic aromatic hydrocarbons in coals and shales from Eastern Nigeria. *Petroleum Coal* 53(3): 188–193.

Olivella, M.À. (2006). Polycyclic aromatic hydrocarbons in rainwater and surface waters of Lake Maggiore, a subalpine lake in northern Italy. *Chemosphere* 63: 116–131.

Ololade, I.A., and Lajide, L. (2010). Exposure level and bioaccumulation of polycyclic aromatic hydrocarbons (PAHs) in edible marine organisms. *J. Environ. Indic.* 5: 69–88. www.environmentalindictorsjournal.net (accessed April 25, 2014).

O'Malley, V.P., Abrajano, T.A., Jr., and Hellou, J. (1996). Stable carbon isotopic apportionment of individual polycyclic aromatic hydrocarbons in St. John's Harbour, Newfoundland. *Environ. Sci. Technol.* 30: 634–639.

Page, D.S., Boehm, P.D., Douglas, G.S., Bence, A.E., Burns, W.A., and Manciewicz, P.J. (1999). Pyrogenic polycyclic aromatic hydrocarbons in sediments record past human activity: A case study in Prince William Sound, Alaska. *Mar. Pollut. Bull.* 38: 247–260.

Pampanin, D.M., and Sydnes, M.O. (2013). Polycyclic aromatic hydrocarbons, a constituent of petroleum: Presence and influence in the aquatic environment. In *Hydrocarbon*, ed. V. Kutcherov. DOI: 10.5772/48176.

Pankow, J.F., Isabelle, L.M., and Asher, W.E. (1984). Trace organic compounds in rain. Sampler design and analysis by adsorption/thermal desorption (ATD). *Environ. Sci. Technol.* 18: 310–318.

Pearlman, R.S., Yalkowsky, S.H., and Banerjee, S. (1984). Water solubilities of polynuclear aromatic and heteroaromatic compounds. *J. Phys. Chem. Data* 13: 555–562.

Pengchai, P., Nakajima, F., and Furumai, H. (2005). Estimation of origins of polycyclic aromatic hydrocarbons in size-fractionated road dust in Tokyo with multivariate analysis. *Water Sci. Technol.* 51(3–4): 169–175.

Pereira, W.E., Hostettler, F.D., Luoma, S.N., van Geen, A., Fuller, C.C., and Anima, R.J. (1999). Sedimentary record of anthropogenic and biogenic polycyclic aromatic hydrocarbons in San Francisco Bay, California. *Mar. Chem.* 64: 99–113.

Pickering, R.W. (1999). A toxicological review of polycyclic aromatic hydrocarbons. *J. Toxicol. Cutan. Ocul.* 18: 101–135.

Plows, F.L., Elsila, J.E., Zare, R.N., and Buseck, P.R. (2003). Evidence that polycyclic aromatic hydrocarbons in two carbonaceous chondrites predate parent-body formation. *Geochim. Cosmochim. Acta* 67(7): 1429–1436.

Poster, D., Schantz, M.M., Sander, L.C., and Wise, S.A. (2006). Analysis of polycyclic aromatic hydrocarbons (PAHs) in environmental samples: A critical review of gas chromatographic (GC) methods. *Anal. Bioanal. Chem.* 386: 859–881.

Prabhukumar, G., and Pagilla, K. (2010). Polycyclic aromatic hydrocarbons in urban runoff sources, sinks and treatment: A review. Department of Civil, Architectural and Environmental Engineering, Illinois Institute of Technology, Chicago. www.drscw.org.

Ravindra, K., Wauters, E., Taygi, S.K., Mor, S., and Van Grieken, R. (2006). Assessment of air quality after the implementation of CNG as fuel in public transport in Delhi, India. *Environ. Monitor. Assess.* 115: 405–417.

Reisen, F., and Arey, J. (2002). Reactions of hydroxyl radicals and ozone with acenaphthrene and acenaphthylene. *Environ. Sci. Technol.* 36: 4302–4311.

Rianawati, E. (2007). Occurrence and distribution of PAHs in rainwater and urban runoff. A thesis submitted for Master of Engineering, National University of Singapore. scholarbank.nus.sg/bitstream/handle/10635/17352/PalaniS.pdf (accessed April 10, 2014).

Ricca, A., Bauschlicher, C.W., Boersma, C., Tielens, A.G.G.M., and Allamandola, L.J. (2012). The infrared spectroscopy of compact polycyclic aromatic hydrocarbons containing up to 384 carbons. *Astrophys. J.* 754: 75–97.

Rivas, F.J., Beltran, F.J., and Acedo, B. (2000). Chemical and photochemical degradation of acenaphthylene: Intermediate identification. *J. Hazard. Mater.* 75: 89–98. http://dx.doi.org/10.1016/S0304-3894(00)00196-5.

RIVM. (2012). *Environmental risk limits for polycyclic aromatic hydrocarbons (PAHs): For direct aquatic, benthic, and terrestrial toxicity.* RIVM Report 607711007/2012. National Institute for Health and Environment (RIVM), Bilthoven, The Netherlands.

Rockne, K.J., Chee-Sandford, J.C., Sanford, R., Hedlund, B.P., Staley, J.T., and Strand, S.E. (2000). Anaerobic naphthalene degradation by microbial pure cultures under nitrate reducing conditions. *Appl. Environ. Microbiol.* 66: 1595–1601.

Ross, R.D., and Crosby, D.G. (1985). Photooxidant activity in natural waters. *Environ. Toxicol. Chem.* 4: 773–778.

Sanders, M., Sivertsen, S., and Scott, G. (2002). Origin and distribution of polycyclic aromatic hydrocarbons in surficial sediments from the Savannah River. *Arch. Environ. Contam. Toxicol.* 43: 438–448.

Schauer, J.J., Kleeman, M.J., Cass, G.R., and Simoneit, B.R.T. (2001). Measurement of emissions from air pollution sources. 3. C1–C29 organic compounds from fireplace combustion of wood. *Environ. Sci. Technol.* 35: 1716–1728.

Senesi, N., and Miano, T.M. (1995). The role of abiotic interactions with humic substances on the environmental impact of organic pollutants. In *Environmental impact of soil component interactions: Land quality, natural and anthropogenic organics,* ed. P.M. Huang, J. Berthelin, J.-M. Bollag, W.B. McGill, and A.L. Page. Vol. 1. CRC Press, Boca Raton, FL, 1995, pp. 311–335.

Shen, H., Huang, Y., Wang, R., Zhu, D., Li, W., Shen, G., Wang, B., Zhang, Y., Chen, Y., Lu, Y., Chen, H., Li, T., Sun, K., Li, B., Wenxin Liu, W., Liu, J., and Tao, S. (2013). Global atmospheric emissions of polycyclic aromatic hydrocarbons from 1960 to 2008 and future predictions. *Environ. Sci. Technol.* 47: 6415–6424.

Sims, R.C., and Overcash, M.R. (1983). Fate of polynuclear aromatic compounds (PNAs) in soil-plant systems. *Residue Rev.* 88: 1–68.

Srogi, Z. (2007). Monitoring of environmental exposure to polycyclic aromatic hydrocarbons: A review. *Environ. Chem. Lett.* 5: 169–195.

Timoney, K.P., and Lee, P. (2011). Polycyclic aromatic hydrocarbons increase in Athabasca River delta sediment: Temporal trends and environmental correlates. *Environ. Sci. Technol.* 45: 4278–4284.

Ukiwe, L.N., Egereonu, U.U., Njoku, P.C., Nwoko, C.I.A., and Allinor, J.I. (2013). Polycyclic aromatic hydrocarbons degradation techniques. *Int. J. Chem.* 5(4): 43–55.

U.S. Department of Health and Human Services. (1995). Toxicological profile for polycyclic aromatic hydrocarbons. www.atsdr.cdc.gov/toxprofiles/tp69.pdf.

USEPA. (1984). Appendix A to Part 136 methods for organic chemical analysis of municipal and industrial wastewater. Method 610—Polynuclear aromatic hydrocarbons. U.S. Environmental Protection Agency, Washington, DC.

USEPA. (1998, July). *Locating and estimating air emissions from sources of polycyclic organic matter.* EPA-454/R-98-014. U.S. Environmental Protection Agency, Office of Air Quality Environmental Protection Planning and Standards, Research Triangle Park, NC.

USEPA. (2001, August). *Emergency planning and community right-to-know act— Section 313: Guidance for reporting toxic chemicals: Polycyclic aromatic compounds category.* EPA 260-B-01-03. U.S. Environmental Protection Agency, Office of Environmental Information, Washington, DC.

USEPA. (2009a). Appendix A to 40 CFR, Part 423—126 priority pollutants. U.S. Environmental Protection Agency, Washington, DC. http://www.epa.gov/NE/npdes/permits/generic/prioritypollutants.pdf (accessed April 14, 2014).

USEPA. (2009b, May). *National primary drinking water regulations.* EPA 816 F-09-004. U.S. Environmental Protection Agency, Washington, DC.

USEPA. (2014, May). *Draft update of human health, ambient water quality criteria, benzo(a)pyrene.* EPA 820-D-14-012. U.S. Environmental Protection Agency, Washington, DC.

Van Metre, P.C., and Mahler, B.J. (2010). Contribution of PAHs from coal-tar pavement sealcoat and other sources to 40 U.S. lakes. *Sci. Total Environ.* 409: 334–344.

Varanasi, U., ed. (1989). *Metabolism of polycyclic aromatic hydrocarbons in the aquatic environment.* CRC Press, Boca Raton, FL, p. 341.

Waller, R.E. (1994). 60 years of chemical carcinogens: Sir Ernest Kennaway in retirement. *J. R. Soc. Med.* 87: 96–97.

Wang, Z., Liu, Z., Xu, K., Mayer, L.M., Zhang, Z., Kolker, A.S., and Wu, W. (2014). Concentrations and sources of polycyclic aromatic hydrocarbons in surface coastal sediments of the northern Gulf of Mexico. *Geochem. Trans.* 15(2). DOI: 10.1186/1467-4866-15-2.

Wärmländer, S.K.T., Sholts, S.B., Erlandson, J.M., Gjerdrum, T., and Westerholm, R. (2011). Could the health decline of prehistoric California Indians be related to exposure to polycyclic aromatic hydrocarbons (PAHs) from natural bitumen? *Environ. Health Perspect.* 119(9): 1203–1207.

Wehrer, M., Rennert, T., Mansfeldt, T., and Totsche, K. (2011). Contaminants at former manufactured gas plants: Sources, properties, and processes. *Crit. Rev. Environ. Sci. Technol.* 41: 1883–1969.

WHO. (1998). *Guidelines for drinking-water quality*, 2nd ed. World Health Organization, Geneva, pp. 123–152.

WHO. (2000). *Air quality guidelines: Polycyclic aromatic hydrocarbons.* 2nd ed. World Health Organization, Regional Office for Europe, Copenhagen, Denmark, chap. 5.9.

WHO. (2008). *Guidelines for drinking water quality.* 3rd ed. World Health Organization, Geneva, 2008.

Wilcke, W., Amelung, W., Martius, C., Garcia, M.V.B., and Zech, W. (2000). Biological sources of polycyclic aromatic hydrocarbons (PAHs) in the Amazonian rain forest. *J. Plant Nutr. Soil Sci.* 163: 27–30.

Wilcke, W., Krauss, M., and Amelung, W. (2002). Carbon isotope signature of polycyclic aromatic hydrocarbons (PAHs): Evidence for different sources in tropical and temperate environments. *Environ. Sci. Technol* 36: 3530–3535.

Wilcke, W., Krauss, M., Lilienfein, J., and Amelung, W. (2004). Polycyclic aromatic hydrocarbon storage in a typical Cerrado of the Brazilian savanna. *J. Environ. Qual.* 33: 946–955.

Wilson, S.C., and Jones, K.C. (1993). Bioremediation of soils contaminated with polynuclear aromatic hydrocarbons (PAHs): A review. *Environ. Pollut.* 81: 229–249.

Yan, J., Wang, L., Fu, P.P., and Yu, H. (2004). Photomutagenicity of 16 polycyclic aromatic hydrocarbons from the USEPA priority pollutant list. *Mutation Res.* 557(1): 99–108.

Yang, C., Wang, Z., Yang, Z., Hollebone, B., Brown, C.E., Landriault, M., and Fieldhouse, B. (2011). Chemical fingerprints of Alberta oil sands and related petroleum products. *Environ. Forensics* 12(2): 173–188.

Yang, H.-H., Lai, S.-O., Hsieh, L.-T., Sueh, H.-J., and Chi, T.-W. (2002). Profiles of PAH emission from steel and iron industries. *Chemosphere* 48: 1061–1074.

Yang, H.H., Lee, W.J., Chen, S.J., and Lai, S.O. (1998). PAH emission from various industrial stacks. *J. Hazard. Mater.* 60: 159–174.

Yu, H. (2002). Environmental carcinogenic polycyclic aromatic hydrocarbons: Photochemistry and phototoxicity. *J. Environ. Sci. Health C Environ. Carcinog. Ecotoxicol. Rev.* 20(2): 149–182.

Yunker, M.M., Macdonald, R.W., Vingarzanc, R., Mitchell, R.H., Goyette, D., and Sylvestrec, S. (2002). PAHs in the Fraser River Basin: A critical appraisal of PAH ratios as indicators of PAH source and composition. *Org. Geochem.* 33: 489–515.

Zafiriou, O.C., Joussot-Dubien, J., Zepp, R.G., and Zika, R.G. (1984). Photochemistry of natural waters. *Environ. Sci. Technol.* 18: 358A–371A.

Zander, M. (1983). Physical and chemical properties of polycyclic aromatic hydrocarbons. In *Handbook of polycyclic aromatic hydrocarbons*, ed. A. Bjorseth. Marcel Dekkel, New York, pp. 1–26.

Zhang, X.L., Tao, S., Liu, W.X., Yang, Y., Zuo, Q., and Liu, S.Z. (2005). Source diagnostics of polycyclic aromatic hydrocarbons based on species ratios: A multimedia approach. *Environ. Sci. Technol.* 39: 9109–9114.

Zhang, Y., and Tao, S. (2009). Global atmospheric emission inventory of polycyclic aromatic hydrocarbons (PAHs) for 2004. *Atmos. Environ.* 43: 812–819.

Zhao, Z.-B., Liu, K., Xie, W., Pan, W.-P., and Riley, J.T. (2000). Soluble polycyclic aromatic hydrocarbons in raw coals. *J. Hazard. Mater.* B73: 77–85.

Zhao, S.M., Wang, B., Wang, D.W., Li, X.M., Huang, B., Hu, P., Zhang, L.W., and Pan, X. (2014). Environmental behavior of PAHs in Dianchi Lake distributions, sources and risk assessment of polycyclic aromatic hydrocarbons in surface sediments from Dianchi Lake, China. *Int. J. Environ. Res.* 8(2): 317–328.

Zhoua, J.L., and Maskaouib, K. (2003). Distribution of polycyclic aromatic hydrocarbons in water and surface sediments from Daya Bay, China. *Environ. Pollut.* 121: 269–281.

3

Quantitative Changes of PAHs in Water and in Wastewater during Treatment Processes

Maria Włodarczyk-Makuła and Agnieszka Popenda

Department of Chemistry, Water and Wastewater Technology, Częstochowa University of Technology, Częstochowa, Poland

CONTENTS

3.1　Water Contamination with PAHs

Anthropogenic activity is regarded as a main source of emission of polycyclic aromatic hydrocarbons (PAHs) into the water environment. This is mainly due to industrial production, e.g., fuel processing, steelworks, and production of coke, creosote, and charcoal, as well as the combustion processes in electrical power plants and waste combustion (Zhou et al. 2005, Park et al. 2009). Among various wastewater types, PAHs occur in the highest concentration in industrial wastewater; however, their presence has been confirmed in municipal waters, runoff waters, and thawing waters. Even wastewater that has been sufficiently treated may discharge significant loads of these pollutants into water environments (Włodarczyk-Makuła 2008b). Moreover, PAHs are leached out of, or washed from, contaminated soils, road surfaces, landfills, and the protection on water supply pipes. They are present in surface waters due to the deposition of solid particles coming from the air during rainfalls. Natural processes are also recognized as sources of PAHs in the environment, including the formation of crude oil, slate, and carbon, and forest and meadow fires. PAHs may also be synthesized by algae and microorganisms. It should be stated that the amount of PAHs introduced into the environment from natural sources is much lower than the amount introduced by anthropogenic sources (Dojlido and Zbieć 1995, Agteren et al. 1998, Ling et al. 2010, Lee and Yoo 2011). The presence of PAHs is also found in ground waters; the contamination of those waters may occur due to the migration of PAHs through soil-water environments. This has important implications for human health, as ground waters are the main source of drinking water (Rejman 1994, Szymański and Siebielska 2000). Runoff waters may infiltrate ground waters, carrying with them PAHs leached from polluted soil, or from soil amended with sewage sludges and compost, as well as leaked from inappropriately sealed landfills. Baran et al. (2004) investigated leaching of PAHs from the mixture of soil and sewage sludges. It was proven that the release of PAHs was possible, despite the fact that sewage sludges originating from the municipal treatment plant were added to soil in dosages that complied with regulations. The infiltration of those compounds through soil layers was proven by comparing the PAHs in various soil layers to those found in the added sewage sludges (Baran et al. 2004). The authors also studied the leaching of PAHs, using the mixture of soil and sewage sludges, as well as sewage sludges coming from municipal and industrial wastewater treatment plants. The experiments were carried out under both static (water extract) and dynamic conditions, including various heights of stored wastes (Twardowska and Włodarczyk-Makuła 1993, Popenda et al. 2002, Włodarczyk-Makuła 2007, Włodarczyk-Makuła and Janosz-Rajczyk 2007, Nowak et al. 2009).

Investigations into contamination of surface and ground waters with PAHs have been conducted for several years as part of the monitoring of water environments. The results of the former studies are often diverse. It is due to the fact that only benzo(a)pyrene (B(a)P) was determined as an indicator of carcinogenic compounds, among either the six PAHs listed by the WHO or the 16 PAHs listed by the USEPA, respectively. The example values of concentrations of those compounds in surface waters in Poland indicate that the total contents of PAHs ranged as high as 19 µg/L. It was demonstrated that the concentration of PAHs depended on the type of compound, place of sampling, and season of the year. A comparison of the results of the studies during dry and wet seasons found that the total concentration of 10 PAHs in dry seasons did not exceed the value of 0.3 µg/L, whereas during rainfalls the concentration could be as high as 16 µg/L (Kowalczyk et al. 2000). The results of the monitoring data originating from the waters of the Vistula and Dunajec Rivers in 2009 are included in Table 3.1. The following PAHs were determined: naphthalene, anthracene, fluoranthene, benzo(a)pyrene, benzo(b)fluoranthene, benzo(k)fluoranthene, indeno(123cd)pyrene, and benzo(ghi)perylene. The sum of the priority concentrations in surface water was below 0.04 µg/L, whereas in ground water the sum ranged as high as 0.54 µg/L (Kułakowski 2009, State of Rivers 2010).

The investigations carried out by the authors examined the water in Goczałkowice Reservoir, which is intended for consumption, for 16 PAHs.

TABLE 3.1

Concentration of PAHs in Waters in Poland, μg/L

PAH	Vistula River	Waters of Basin of Dunajec River	
		Surface Waters	**Ground Waters**
Naphthalene	0.0025–0.035	Below 0.005	
Anthracene	0.003–0.01	Below 0.005	
Fluoranthene	0.007–0.0197	Below 0.018	
Benzo(a)pyrene	0.001–0.0043	Below 0.012	0.16
Benzo(b)fluoranthene Benzo(k)fluoranthene	0.002–0.015	Below 0.005	0.06–0.09
Benzo(ghi)perylene	0.002–0.012	Below 0.003	0.11
Indeno(123cd)pyrene		Below 0.005	0.13

Source: State of Rivers Based on the Results of Investigations Carried Out in the Frame on National Environmental Monitoring in Years 2007–2009, Library of Environmental Monitoring, Main Inspector of Environmental Protection, Warszawa, 2010 (in Polish); Kułakowski, Priority Substances and Other Contaminants in Water Environment, 2009, www.wios.tarnow.pl (in Polish).

The total concentration of PAHs was equal to 0.0096 μg/L. Four-ring hydrocarbons were dominant, mainly benzo(b)fluoranthene and benzo(k) fluoranthene. The total concentration of the aforementioned hydrocarbons did not exceed the average yearly values in the regulations pertaining to environmental quality standards for priority compounds (0.03 μg/L). The concentration of benzo(a)pyrene ranged up to 0.008 μg/L, but did not exceed the permissible value determined in environmental quality standards for uniform bodies of surface waters, which is 0.1 μg/L (Nowacka and Włodarczyk-Makuła 2013). The toxicological investigations described in the literature have repeatedly shown the toxic impact of PAHs on water organisms. They have carcinogenic and mutagenic properties. There is also the possibility of PAHs negatively influencing reproduction, as well as initiating hereditary genetic and ecotoxic defects (Augusto et al. 2008, Czarnomski and Izak 2008, Martorell et al. 2010). The laboratory studies indicated that PAHs are not toxic themselves, but that their disadvantageous impact is related to their metabolites. These are products of fates of internal cells' nonsubstituted PAHs. The aforementioned fates take place under microsomal enzymes bounded to cytochrome P-450. Much of the data in the literature give results of individual hydrocarbons in relation to bacteria and fungi, as well as indicators occurring as individual colonies or species. Among PAHs, benzo(a)pyrene is thought to have the most significant impact on indicator organisms. However, five-ring and six-ring hydrocarbons have similar properties. In Table 3.2 the activity of selected hydrocarbons together with carcinogenic doses for experimental animals is presented.

Data in the literature indicate that the PAH-related compounds formed in reactions with other compounds of the environment are often more toxic

TABLE 3.2

PAH Activity

PAH	Carcinogenity	Carcinogenity Dose for Experimental Animals, µg/kg	Mutagenicity	Relative Carcinogenic Coefficient
Benzo(a)pyrene	++++	2	++++	1
Dibenzo(ah)anthracene	++	6	++	5
Benzo(a)anthracene	+	2000	++	0.1
Chrysene	+	99,000	++	0.01
Benzo(b)fluoranthene	+++	40,000	++	0.1
Benzo(k)fluoranthene	+	72,000	++	0.1
Benzo(ghi)perylene	+++		++	0.01
IP	+	72,000	+	0.1

Source: Data from Agteren et al., *Handbook on Biodegradation and Biological Treatment of Hazardous Organic Compounds*, Kluwer Academic Publishers, Dordrecht, The Netherlands, 1998.

Note: +, weak activity; ++, average activity; +++, strong activity; ++++, very strong activity.

than those that are not substituted. Carcinogenic and mutagenic activity has been proven for nitrite-related compounds such as 1-nitropyrenem 3-nitrofluoranthene, 2,3-dinitrotoluene, and 2,6-dinitrotoluene formed as a result of UV radiation, ozone nitrogen oxide (V), and PAH-related halogens, including sulfur (Piekarska 2008). There are also listed substances that are suspected to be carcinogenic for humans (group 2B according to the International Agency for Research on Cancer (IARC)). In in vitro studies carried out on bacteria and mammal cells (including humans), nitro compounds indicated genotoxic activities. In vivo and in vitro studies have demonstrated mutagenic and clastogenic activity of nitrite metabolites of PAHs (Makhniashvili 2003). Grung and others investigated the identification of PAH metabolites absorbed into fish organisms. Fish were exposed to the presence of pollutants in water, as well as in feed. Monohydroxy metabolites were found in the bile of the fish. It was concluded that the investigated metabolites of PAHs are reliable markers of the exposure of fish to hydrocarbons. It was also shown that there is a relation between metabolites present in the bile of fish studied and concentrations of PAHs in water and feed, respectively (Grung et al. 2009, Harman et al. 2009).

The results of toxicological investigations described in the literature indicate significant variety, and the LC50 values differ by as much as three orders of magnitude. An example of this are the studies carried out by Kalinowski and Załęska-Radziwiłł (2005), whose goal was to estimate the toxicity of selected PAHs on water organisms using survival tests, enzymatic tests, and growth tests. Naphthalene was found to be highly toxic to shellfish, whereas phenanthrene and pyrene were highly toxic to algae, shellfish,

TABLE 3.3

Lethal Concentration with Respect to PAHs

Test Organism	Time of Test (h)	LC(EC) 50 t, mg/L			
		Naphtalene	Phenanthrene	Anthracene	Pyrene
Fish (*Lebistes reticulatus*)	96	138.2	40.5	141.4	176.8
Algae (*Selenastrum capricornutum*)	72	47.1	0.27	0.44	0.39
Shellfish (*Daphnia magna*)					
Survival test	48	101.8	75.5	44.6	60.5
Shellfishes (*Daphnia magna*)					
Enzymatic test	1	0.59	0.96	3.6	0.86
Bacteria (*Vibrio fisheri*)					
Survival test	0.25	21.9	3.32	0.61	2.25
Bacteria (*Vibrio fisheri*)					
Survival test	0.5	25.41	2.90	0.65	2.43
Oligochaeta (*Tubifex tubifex*)	48	166.7	191.3	15.7	156.6
Larvae (*Chironomus*)	48	75.0	56.3	6.4	197.7

Source: Data from Kalinowski R., and Załęska-Radziwiłł M., Assessment of PAH toxicity toward freshwater organisms, presented at Conference on Micropollutants in the Human Environment, Conference 56, Czestochowa University of Technology, Czestochowa, 2005.

and bacteria. Anthracene was highly toxic to algae and bacteria. Based on the investigations, the safe (acceptable) concentrations of the hydrocarbons in the study were determined. The highest safe concentration determined, that of naphthalene, was equal to 1.248 µg/L. In the case of anthracene, the safe concentration was on the level of 0.105 µg/L; the lowest safe concentrations determined were for phenenthrene and pyrene, at 0.031 and 0.015 µg/L, respectively. The results of these studies are included in Table 3.3.

3.2 Determination of PAHs in Water and Wastewater

There have been over 300 PAHs identified in the environment. However, in scientific research or monitoring projects for individual environmental elements, only a selected group of compounds—e.g., 6, 8, 9, 13, 15, 16, or 18—are determined (Dz. Urz. UE seria L 158, 2004; Dz. U Nr 165 poz. 1359, 2002; Dz. U Nr 72 poz. 466, 2010; Dz. U Nr 254 poz. 1528, 2011; Dz. U Nr 258 poz. 1550, 2011; Dz. U Nr 162 poz. 1008, 2008; Dz. U Nr 298 poz. 1771, 2011; PN-EN ISO 17993:2005; Method 8310).

The U.S. Environmental Protection Agency (EPA) lists 16 PAHs that should be analyzed in environmental samples:

Naphthalene
Acenaphythlene
Acenaphtene
Phenanthrene
Anthracene
Fluorene
Pyrene
Fluoranthene
Benzo(a)anthracene
Chrysene
Benzo(b)fluoranthene
Benzo(k)fluoranthene
Benzo(a)pyrene
Dibenzo(ah)anthracene
Benzo(ghi)perylene
Indeno(123cd)pyrene

This list originates as an extension of the World Health Organization list, which is comprised of the following six hydrocarbons:

Fluoranthene
Benzo(b)fluoranthene
Benzo(k)fluoranthene
Benzo(a)pyrene
Benzo(ghi)perylene
Indeno(123cd)pyrene

Legislation of the European Union obliges the owners of installations emitting PAHs into the environment to provide information to the National Register of the Release and Transfer of Contaminants. The maximum allowable load of the aforementioned contaminants' discharges into waters and soils is at the level of 5 kg per year (Dz. Urz. WE L33/2006). There are eight PAHs listed in the Directive of the European Parliament and Council on the Environmental Quality Standards List:

Anthracene
Naphthalene

Fluoranthene

Benzo(b)fluoranthene

Benzo(k)fluoranthene

Benzo(a)pyrene

Benzo(ghi)perylene

Indeno(123cd)pyrene

The same compounds are specified in the register of priority substances, and six of them (the exceptions being naphthalene and fluoranthene) are regarded as priority dangerous (Dz. Urz. 348//2008).

In Poland, the regulation for the method of classifying the status of uniform bodies of surface waters gives the maximum allowable values for eight selected PAHs, which are regarded as priority for the water environment (Dz. U Nr 257 poz. 1545, 2011). The same eight PAHs are listed in regulations on the method and ways of conducting diagnostic monitoring of surface waters (Dz. U Nr 258 poz. 1550, 2011). The border levels of PAHs are also given in the current regulation regarding requirements of surface waters intended for consumption (categories of quality of water A1, A2, A3). The allowable concentration for A1 and A2 categories is set at the level of 0.2 µg/L, whereas for category A3, at the level of 1.0 µg/L (Dz. U Nr 204 poz. 1728, 2002). In the Ministry of Health's regulation on the quality of water intended for human consumption, there are five PAHs listed. The allowable concentration for benzo(a)pyrene is set at the level of 10 µg/L, while the total allowable concentration of the four hydrocarbons—benzo(b)fluoranthene, benzo(k)fluoranthene, benzo(ghi)perylene, and indeno(123cd)pyrene—is set at the level of 100 ng/L (Dz. U Nr 72 poz. 466, 2010). In order to estimate the state of ground waters, B(a)P and the sum of PAHs are given. The range of the concentration of B(a)P for hydrogeochemical background is 1–10 ng/L. Allowable concentrations for classes I, II, III, and IV are 10, 20, 30, and 50 ng/L, respectively. If the concentration is higher than 50 ng/L, then it is regarded as class V. The hydrogeochemical background for PAHs other than B(a)P is given in the range of 1–100 ng/L. For individual classes from I to IV, the border concentrations are equal to 100, 200, 300, and 500 ng/L, respectively. If the concentration is above the value of 500 ng/L, then the waters are regarded as class V (Dz. U Nr 143 poz. 896, 2008). In the Ministry of the Environment's regulation on the conditions that should be fulfilled for wastewater discharge into waters or soils, PAHs are listed as compounds that require elimination due to their carcinogenic, mutagenic, and teratogenic properties (Dz. U Nr 27 poz. 169, 2009).

In 1985 Her Majesty's Stationery Office (HMSO) of Great Britain published a document describing the method for determination of PAHs in aquatic environments. Based on this procedure, the EPA method (Methods 8100 and 8310) was devised, utilizing high-performance liquid chromatography (HPLC) in order to quantify the PAHs. These methods describe the procedures for preparation of samples on the basis of liquid-liquid extraction, continuous extraction, and extraction to the solid phase. As chromatographic techniques developed, guidelines could be elaborated for the determination of PAHs using capillary gas chromatography with mass spectrometry (GC-MS) and the application of deuterated internal standards. EPA Method 610 is designed for the determination of PAHs in municipal and industrial wastewater, based on the extraction of dichloromethane and quantification using HPLC or GC with packed columns. In 2002, within the framework of standards of water quality, ISO 17993:2005 was designed, which addressed the analysis of 15 PAHs in surface waters and ground waters intended for consumption; it also can be adjusted for wastewater analyses. It necessitated the use of dark/brown glass, stabilization with thiosulfate sodium, and extraction with hexane (25 ml to the sample of 1 L) and drying with sulfate sodium and a concentration of extract in an evaporator. HPLC with fluorescence detector is recommended for identifying PAHs (Wolska 2008). There are several steps in order to identify and qualify PAHs that impacted the precision of the results; these are shown in Figure 3.1 (Namieśnik et al. 2000, Grajek 2003, Gdaniec-Pietryka et al. 2007, Rosik-Dulewska 2007, Mechlińska et al. 2010b).

In the case of environmental samples, the significant step is the sampling and preparation of selected representative samples into determination. The prestep

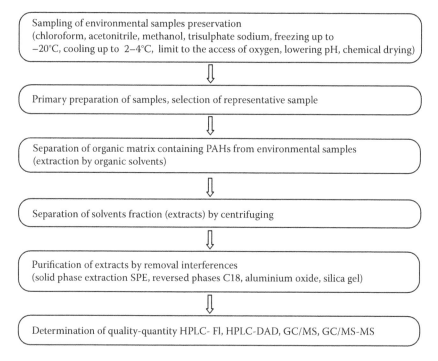

FIGURE 3.1
General scheme of preparation of environmental samples in PAH determination.

is isolation of organic matter, including, among others, PAHs. Liquid-liquid extraction is commonly used in water or wastewater. Other extraction methods listed by the EPA and applied by various authors are as follows:

Soxhlet extraction (Method 3540C) and automatic extraction in Soxhlet apparatus (Method 3541)

Continuous liquid-liquid extraction (Method 3520C)

Supercritical liquid extraction SFE (Method 3561)

Extraction supported with microwave radiation MAE (Method 3546)

Ultrasonic extraction (Method 3550C)

Pressurized fluid extraction PFE (Method 3545A)

The next step in sample preparation is selection of analyzed compounds from the interferences isolated from the samples during extraction. The following processes are applied in order to isolate PAHs from solvent extracts:

Purification of silica gel (Method 3630C)

Purification on oxide aluminum (Method 3611B)

Purification of gel-permeate GPC (Method 3640A)

Purification of acid-alkaline (Method 3650B)

Sulfuric acid/permanganate cleanup (Method 3665A)

The final step in the determination of PAHs is the analysis of their quality and quantity using chromatographic methods. HPLC and GC utilizing various detectors are commonly applied. HPLC is given as a reference method for determination of PAHs in the Polish Ministry of the Environment's legislation on conditions that should be fulfilled for wastewater discharge into waters and soils. The PN-EN ISO 17993 standard regarding determination of 15 PAHs in water using the HPLC method with fluorescence detection after liquid-liquid extraction was elaborated. The isolation of PAHs from environmental samples is carried out as follows:

Shaking with the solvent

Sonification

Extraction with the usage of gases under an overcritical state

Continuous extraction or shaking with the solvent

Cyclohexane, dichloromethane, hexane, and acetone are commonly used solvents (PN-EN ISO 17993). However, the isolation ability of detectors applied with HPLC is lower than those applied with GC-MS or GC-MS/MS, respectively. The application of a quadrupole detector together with gas chromatography allows for a lower threshold of determination, and improves the selectivity as well as identification of quantity and quality of compounds occurring in trace amounts.

There are many publications that examine the analysis of PAHs. The procedures of preparation of environmental samples (surface waters and sediments, ground waters, air, soil) and many other liquid and solid samples (e.g., wastewater, sewage sludges, supernatants, and wastes) for chromatographic analysis are numerous and varied. Therefore, an established practice during investigations is to verify the method used for determining the recovery of PAHs, by adding the standard mixture, certified materials, and hydrocarbons determined by the atoms of isotope of carbon or hydrogen. There are various mixtures of solvents, methods of extraction, and chromatographic techniques. For example, Wolska carried out the comparison of recovery of PAHs from water and application of certified material of sediments. The investigations were carried out using certified material—sediments, parallel in three laboratories. The recoveries were determined for eight hydrocarbons. Three different methods were used: liquid-liquid with added hexane, extraction on disc C18 (elution of dichloromethane), and liquid-solid extraction on cellulose filter (isopropanol). Table 3.4 shows the ranges of recoveries of the PAHs studied (Perez et al. 2001, Grajek 2003, Wolska et al. 2005, Hua et al. 2008, Pena et al. 2008, Wolska et al. 2008, Wolska et al. 2009).

Mechlińska and coauthors (2010b) compiled the data in the literature concerning recovery values of PAHs from water and sediment samples, using hydrogen isotope standards. The obtained values ranged from 10 to 119%.

TABLE 3.4

Ranges of Recoveries (%) of PAHs from Water

PAHs	Range Values	PAHs	Range Values
Naphtalene	16–95	Pyrene $^2H_{10}$	85
Acenaphtalene	17–100	Fluorantene	9–114
Acenaphtene $^2H_{10}$	65–80	Chrysene $^2H_{12}$	63–82
Fluorene	19–94	Benzo(a)anthracene	8–103
Phenanthrene	64–97	Dibenzo(ah)anthracene	10–97
Phenanthrene $^2H_{10}$	72	Benzo(a)pyrene	14–99
Benzo(b)fluoranthene	27–101	Benzo(ghi)perylene	11–107
Benzo(k)fluoranthene	22–100	Indeno(123cd)pyrene	11–101
Pyrene	11–99	Anthracene	17–102

Source: Wolska L., et al., *Analytical and Bioanalytical Chemistry*, 382, 2005, 1389–1397; Hua L., et al., *Biomedical and Environmental Science*, 21, 2008, 345–352; Grajek H., Factors Influencing on the Effectiveness and Selectivity of Extraction to the Solid Phase, presented at Conference on the Micropollutants in the Human Environment, Conference 51, Czestochowa University of Technology, Czestochowa, 2003, 347–371 (in Polish); Wolska L., *Analytical and Bioanalytical Chemistry*, 391, 2008, 2647–2652; Helaleh M.I.H., et al., *Journal of Chromatography A*, 1083, 2005, 153–160; Perez S., et al., *Journal of Chromatography*, 938, 2001, 57–65.

The method of internal standard applying hydrocarbons determined with the following isotope of hydrogen was used: $^2H_8$ (naphtalene), $^2H_{10}$ (acenaphtene, phenanthrene, fluoranthene, anthracene), and $^2H_{12}$ (chrysene, benzo(a) anthracene, benzo(a)pyrene, benzo(ghi)perylene), respectively. In the case of the standard mixture added to the water as an internal standard, the range was narrow as the recoveries were in the range of 70 to 92% for four hydrocarbons: naphtalene $^2H_8$, acenaphtalene $^2H_{10}$, phenanthrene $^2H_{10}$, and chrysene $^2H_{12}$ (Mechlińska et al. 2010a).

The authors' investigations in the analysis of PAHs were carried out using samples of surface water, municipal and industrial wastewater, and supernatants. In order to determine the recovery, the standard mixture of 16 compounds recommended by the EPA for environmental analyses was used. To analyze PAHs, the preparation of water samples was carried out by liquid-liquid extraction (LLE) using hexane as an organic solvent. Extraction was conducted using a horizontal shaker for 60 min. Then the extracts were concentrated to the volume of 2 ml under nitrogen stream. The quantification of 16 PAHs was carried out using a gas chromatograph coupled with mass spectrometer, GC800/MS800. Interpretation of the results was obtained by comparing the retention times and spectra mass of the hydrocarbons present in the sample to the retention times and standard mixture spectra (Nowacka and Włodarczyk-Makuła 2013). Investigations into the identification of PAHs in raw and treated municipal and industrial wastewater, as well as in supernatants originating from various sewage sludges (mineral, raw, and digested), were carried out. Figure 3.2 gives an overview of the preparation of wastewater or supernatant samples for PAH analysis.

Factors taken into consideration included the type of solvents and mixing (liquid-liquid shaking, sonification), the amount of added standard mixture, and the method for purifying extracts (silica gel, cyanopropyl) (Włodarczyk-Makuła 2007, 2008a). Table 3.5 lists the ranges of recovery values for the samples studied, of wastewater and supernatants separated from sewage sludges (Włodarczyk-Makuła 2007, 2008a).

The recovery range of individual hydrocarbons is diverse, due to the complexity of the organic matrix in wastewater and supernatants. The recovery values are higher for distilled water with the standard mixture of PAHs added; they vary from several percent to 120%. In wastewater the presence and chemical characteristics of suspensions are of great importance. The separation of suspensions by filtration (including gel or membrane) and centrifuging is often the reason for high losses of PAHs, due to their sorption onto solid, mainly organic, particles. The results of investigations available in the literature indicate the nonuniform way of both sample preparation and determination of their quality and quantity. Thus, the results are often diverse and difficult to compare. This is the consequence of the heterogeneity of the environmental matrix, variations in extraction efficiency, as well as differences in accuracy, detection limits, and precision of analytical facilities/systems (chromatographic systems).

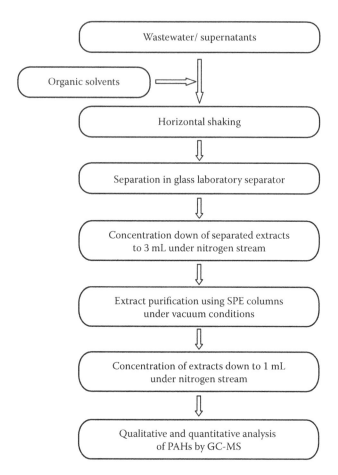

FIGURE 3.2
Preparation of wastewater samples for determination of PAHs by GC-MS.

3.3 Changes in Concentration of PAHs in Water and Wastewater during Treatment Processes

Although PAHs are regarded as persistent organic pollutants occurring in the environment, they undergo changes (fates) as conditions change or various factors occur. The fates of PAHs are related to biodegradation and transformation without the presence of microorganisms. The abiotic processes are the following: vaporization, chemical oxidation, sorption, photodegradation, and reactions with other chemical compounds (substitute and additive reactions). The migration of PAHs into other elements of the environment occurs due to vaporization and sorption. Degradation of PAHs is possible as a result of photodegradation and chemical reactions. However, during these

TABLE 3.5

Ranges of Recovery (%) of PAHs from Wastewater and Supernatants

	Range Values			Range Values	
PAH	**Wastewater**	**Supernatants**	**PAH**	**Wastewater**	**Supernatants**
Naphtalene	30–74	30–52	Benzo(a) anthracene	51–95	48–83
Acenaphtalene	35–108	37–73	Chrysene	50–97	46–84
Acenaphthene	34–106	40–87	Benzo(b) fluoranthene	49–81	39–95
Fluorene	36–114	41–82	Benzo(k) fluoranthene	43–115	41–92
Phenanthrene	39–108	63–108	Dibenzo(ah) anthracene	38–87	30–81
Anthracene	40–119	53–93	Benzo(a) pyrene	47–97	55–92
Fluoranthene	46–117	38–85	Benzo(ghi) perylene	34–81	60–75
Pyrene	45–107	45–82	Indeno(123cd) pyrene	48–75	45–82

Source: Włodarczyk-Makuła M., Quantitive Changes of PAHs during Wastewater and Sludges Treatment, Monograph 126, Wydawnictwo Politechniki Częstochowskiej, Częstochowa, 2007 (in Polish); Włodarczyk-Makuła M., *Archives of Environmental Protection*, 3, 2008a, 259–264.

processes substitute compounds are formed, whose identification is not completed at present, and their toxicity may be higher than that of the original compounds (Jamróz et al. 2002, Włodarczyk-Makuła 2011a).

In water purification, PAHs are removed during filtration, infiltration, and chlorination processes. It is considered that biodegradation plays a significant role in the decrease of PAH concentrations during the process of slow filtration. During the process of infiltration, adsorption and bioaccumulation are of prime importance. The quick filtration together with chlorination allows for efficient removal of PAHs. Furthermore, in the process, chlorine substitutions of high toxicity occur. They are quinones, epoxide compounds, methylene, and carboxyles. The advance oxidation processes are efficient in the removal of PAHs. Ozone can be applied to purify the water, given proper contact times and doses; this can ensure the total removal of PAHs. In the case of chemical oxidation, it is necessary to apply activated carbon as a further process, to ensure the removal of bypass products of volatilization. When using membrane processes, chemical oxidation, or adsorption of activated carbon, it is important to remove suspensions and organic compounds. This allows the removal of PAHs that occur as sorption compounds. If PAHs occur in the water intended for consumption, then the conventional processes are not sufficient to remove these compounds completely (Traczewska 2003). This is confirmed by the results of the authors' investigations. The concentration of PAHs was determined in the water originating from the surface intake, as well as in water coming from the treatment processes. The following samples were taken: raw water, after primary ozonation, and coagulation with sulfate (VI) aluminum, after sedimentation, after filtration on quick sand filters, after ozonation, after adsorption on active carbon, and after final chlorine disinfection. The total concentration of 16 PAHs in raw water was equal to 96 ng/L. The concentration of PAHs in the water stream, after each stop of the purification process, varied from 39 to 203 ng/L. The significantly lower concentration of PAHs was found in the water after the coagulation, sedimentation, and disinfection processes, whereas periodically higher concentrations of PAHs were found after filtration. The final total concentration of 16 analyzed PAHs was lower than the initial concentration of 15%. Therefore, in the water treatment plant, changes of the total 16 PAHs (comparing initial and final) were not statistically significant (Nowacka and Włodarczyk-Makuła 2013).

As mentioned, PAHs undergo the process of photodegradation due to their sensitivity to electromagnetic radiation: gamma radiation and ultraviolet (Cataldo and Keheyan 2006). Thus, photodegradation and oxidation, and their combination, are regarded as the most efficient methods for the destruction of PAHs (Dugay et al. 2002, Jamróz et al. 2002). The impact of photodegradation on PAHs as a function of the concentration of fulvic acids in waters was investigated by Xia et al. (2009). When ranked by susceptibility to photolysis, the hydrocarbons were in the following order: acenaphytelene > pyrene > phenanthrene > fluorene > fluoranthene.

The level of degradation of Ace, Flu, and Fen decreased when the concentration of fluvic acids increased. In the case of pyrene and fluoranthene, the efficiency of photolysis was higher at low acid concentrations. It was shown that the presence of acids alters the formation of active oxygen, and therefore the level of degradation of PAHs was twice as high (Xia et al. 2009). Cataldo and Keheyan (2006) investigated the effectiveness of gamma radiation on the removal of 17 PAHs dissolved in acetonitrile. Coronene was found to be the most persistent. The persistence of the hydrocarbons investigated was as follows: coronene > chrysene > fluorine > 2-metylophenanthrene > acenaphylene > pyrene > 3-thylophenanthrene > fluoranthene > phenanthrene > 1-methylonaphtalene > anthracene > naphthalene.

The total removal of PAHs was possible after the application of additional oxidation using ozone. A higher dose of radiation (200 kGy) was proposed as an alternative solution. Total removal was obtained of fluorine and chrysene from acetonitrile. PAHs may undergo the process of degradation under ozone, but the products of the radiolysis, as well as radiolysis together with ozonation, will be varied. It is due to this fact that PAH-related compounds that are formed after exposition to irradiation using ozone undergo the process of oxidation (Cataldo and Keheyan 2006).

Ferrarese et al. (2008) investigated the susceptibility of PAHs to oxidation using strong oxidants. The studies were carried out using dioxide dihydrogen, Fenton reagent, activated sulfate sodium, permanganate, and a combination of reagents, with the initial concentration of PAHs equal to 2800 mg/g d.w. The highest efficiency of oxidation seen was 95%; this was obtained during oxidation with Fenton reagent (Ferrarese et al. 2008). In the process of sorption, an accumulation of PAHs in sewage sludges or on sorbents takes place. Most of the results reported in the literature concern estimating the effectiveness of mineral oils' sorption on active carbon or other sorbents. In the case of PAHs, there are limited data in the literature. For example, the kinetics of sorption of PAHs in seawater was determined based on the results originating from the application of sorbents of polypropylene or caoutchouc. It was proven that using caoutchouc sorbent, the effectiveness of the removal of studied PAHs was higher than that of those with polypropylene (Yakan et al. 2009).

The authors carried out investigations monitoring changes in the concentration of PAHs in wastewater during treatment. The studies concerned treatment in the municipal treatment plant (mechanically biological with the chemical removal of phosphorus). The concentrations of PAHs in raw wastewater, after sand traps, after primary settle tanks, after biological reactors, and in the treated wastewater were determined, respectively. It was proven that in raw wastewater, the concentration of 16 PAHs varied from 4.8 to 6.5 µg/L. During the process of sedimentation of mineral suspension solids, a decrease of 11 to 12% occurred. The analysis of PAHs selected from wastewater during this process indicated the high concentration of PAHs (450–642 µg/kg d.w.). After the process of sedimentation in primary settling tanks, the further decrease of PAHs in wastewater took place, and their sorption/cumulation in sewage sludges was observed. The total concentration of PAHs was lower by 40 to 47% than that of the initial contents. In the treated wastewater, the concentration of 16 PAHs was lowered by 78 to 93% from that of the raw wastewater. Taking into account the amount of treated wastewater, the load of PAH discharges may be in the range of several dozen kilograms per year (including about 30% carcinogenic PAHs). The high concentrations of PAHs found in sewage sludges demonstrated the affinity of these compounds for sorption onto solid matters under conditions of slow flows (Włodarczyk-Makuła 2007). Because the processes applied in the treatment plant did not remove PAHs sufficiently, further investigations were conducted in order to treat wastewater after the biological processes. The aim was to remove PAHs from wastewater using the oxidation process. The changes in the concentration of PAHs in biologically treated municipal and industrial wastewater with the usage of dioxide dihydrogen were analyzed (Włodarczyk-Makuła 2011b, Turek and Włodarczyk-Makuła 2012, Włodarczyk-Makuła and Turek 2012). In the investigations of municipal wastewater, changes in the concentration of PAHs were analyzed in both wastewater samples taken from the wastewater treatment plant and those with the standard mixture added. The decrease of the total concentration of PAHs in wastewater with the added PAHs ranged from 65 to 71%. This was higher than the decrease seen in compounds in wastewater without the standard mixture added; there the decrease did not exceed 59% (range 51 to 59%). The investigations into the removal of PAHs from wastewater were also carried out in industrial wastewater. The wastewater discharge from coking plant processes was selected due to their high loads of PAHs. The percentage removal of individual hydrocarbons depended on the dose of dihydrogen dioxide, and the type of hydrocarbon, and ranged as high as 86%.

The other investigations into the removal of PAHs from wastewater primarily treated in municipal treatment plants examined the process of photodegradation with UV radiation. The initial concentration was in the range of 0.8 to 1.2 µg/L. The samples of wastewater (with and without the standard mixture added) were exposed to UV radiation of 254 nm wavelength. It was found that the initial contents of PAH impacted the effectiveness of their removal. In wastewater without the standard mixture added, the effectiveness of the removal of PAHs ranged up to 65%, whereas in that with the standard mixture added, the effectiveness of the removal of PAHs ranged up to 84% (Włodarczyk-Makuła 2011a).

The data in the literature indicate that membrane processes can also be applied to remove organic micropollutants, including PAHs, in the processes of purification of water. Pressure processes are regarded as the most useful, due to the size of the particles. Among them, nanofiltration is the most effective process in the removal of micropollutants, which are colloids, low-particle organic compounds, or two-valence ions. For five hydrocarbons the values of retention coefficient varied from 39 to 96% and from 45 to 99%, for reversed osmosis (RO) membranes and nanofiltration (NF) membranes, respectively. It was shown that ultrafiltration may also be successfully applied in order to separate PAHs from water solutions, despite the fact that molecular weights are lower than the border selectivity cutoff and low in view of the pore radius of membranes applied in the aforementioned process (Jin et al. 2005, Bodzek and Konieczny 2010). A comparison of the effectiveness of the removal of PAHs from water in the processes of ultrafiltration and nanofiltration, as well as reversed osmosis, was carried out by Dudziak et al. (2003). The determined retention coefficients of PAHs originating from the model solutions were in the range of 15 to 99% (Dudziak et al. 2003, Dymaczewski and Jeż-Walkowiak 2012). The application of ultrafiltration (UF) and reversed osmosis in removal of PAHs from industrial wastewater as well as leachates coming from landfills was studied by the authors (Smol and Włodarczyk-Makuła 2012a, 2012b). The changes in the concentration of PAHs in coke plant wastewater in UF with primarily filtration were analyzed. The biologically treated wastewater originating from the wastewater treatment plant was used in the study. The results using primarily filtration were: the total level of removal of PAHs ranged up to 85%, whereas for individual hydrocarbons, removals in the range of 27 to 100% were seen (Smol and Włodarczyk-Makuła 2012a). When investigating leachates, reversed osmosis was applied. The percentage decrease of PAHs in the studied processes ranged as high as 86%. For individual compounds, the efficiency of the removal ranged from 42 to 100% (Smol and Włodarczyk-Makuła 2012b).

Biodegradation in water environments may take place with the aid of bacteria, fungi, and some algae. The biological fates of PAHs are related to bioaccumulation as well as biodegradation (Librando et al. 2003). The biological fates may be carried out by individual strains of bacteria or by the set of microorganisms living in symbiosis. The microorganisms present in the environment are usually not able to decompose these compounds if the environment becomes polluted. A certain period of time is needed for adaptation, to bring forth the ability to produce the proper enzymes—directly or after some genetic changes—as a result of the entrance of PAHs in metabolic pathways. The fates may take place under both aerobic and anaerobic conditions; however, their intensity is higher in the presence of oxygen (Bernal-Martinez et al. 2007).

It is stated in the literature that the hydrocarbons best able to biodegrade are those of two and three benzene rings. PAHs may be the sole source of carbon for bacteria, but the source is not easily available. In the presence of other easily available carbon sources, co-metabolism or the decomposition of many compounds is possible. The first step in the process of biodegradation of PAHs is the oxidation of benzene rings into cis-dihydrodiols in the presence of monooxygenase. Dioxygenase is the catalyst for these reactions. Then, cis-dihydrodiols are converted to catechols; of those, the fates proceed in two directions: ortho and meta pathways. In the ortho pathway there is a decomposition among carbon atoms that are substituted for hydroxyl (–OH) groups that results in forming cis,cis-muconic acid, and then acetylo- and succiniclocoenzyme A. In the fates on the meta pathway, bonds are broken between carbon atoms; of these, one is substituted for a hydroxylic group. Then, hydroxymuconic semialdehydes are formed; formic aldehyde and then pyruvate acid are formed, respectively. The final products of the decomposition on the pathways may be entered into metabolic fates taking place in cells (Krebs cycle) (Traczewska 2003, Haritash and Kaushik 2009).

The efficiency of biodegradation may also be stimulated chemically. Environmental conditions such as temperature, pH, and water content impact the biodegradation efficiency. The process of biodegradation of PAHs under aerobic conditions was investigated by Zhao et al. (2006). The removal of PAHs was in the range of 84 to 90%, depending on the strain of bacteria (Zhao et al. 2006).

Załęska-Radziwiłł et al. (2007) studied the process of biodegradation of the following PAHs: naphthalene, phenanthrene, anthracene, and pyrene. The biodegradation test was conducted according to the OECD MITI(I)-301C method. The experiments were carried out under aerobic conditions. The initial concentration of hydrocarbons was equal to 100 mg/L; after 14 days of incubation, the concentration of naphthalene and pyrene was lower by 99%. In the case of anthracene and phenanthrene, reductions of 97 and 98% were obtained after 28 days. The toxicity tests confirmed the formation of toxic metabolites, especially in the biodegradation of phenanthrene (Załęska-Radziwiłł et al. 2007).

References

Scientific Literature

Agteren M.H., Keuning S., Janssen D.B., *Handbook on biodegradation and biological treatment of hazardous organic compounds*, Kluwer Academic Publishers, Dordrecht, The Netherlands, 1998.

Augusto S., Goncalves M., Maguas C., Branquinho C., Mendes B., Determination of polycyclic aromatic hydrocarbons in chicken meat, liver and eggs, presented at International Symposium on Green Chemistry for Environment and Health, Monachium, 2008.

Baran S., Bielińska J., Oleszczuk P., Enzymatic activity in an air field soil polluted with polycyclic aromatic hydrocarbons, *Geoderma*, 118, 2004, 221–232.

Bernal-Martinez A., Carrere H., Patureau D., Delgenes J.-P., Ozone pre-treatment as improver of PAH removal during anaerobic digestion of urban sludge, *Chemosphere*, 68, 2007, 1013–1019.

Bodzek M., Konieczny K., The usage of membrane techniques in purification of waters intended for drinking purposes. Part II. The removal of organic compounds, *Water Technology*, 2, 2010, 15–31 (in Polish).

Cataldo F., Keheyan Y., Gamma-radiolysis and ozonolysis of polycyclic aromatic hydrocarbons (PAHs) in solution, *Journal of Radioanalytical and Nuclear Chemistry*, 267, 3, 2006, 679–683.

Czarnomski K., Izak E., Persistent organic contaminants in the environment, Legislation of European Community Nr 850/2004, Information Materials, Ministry of Environment, Institute of Environmental Protection, Warsaw, 2008 (in Polish).

Dojlido J., Zbieć E., Organic pollutants in Warsaw drinking water, *Environmental Protection*, 3, 1995, 55–58 (in Polish).

Dudziak M., Bodzek M., Luks-Betlej K., The removal of PAHs from waters using membrane processes, *Engineering and Protection of Environment*, 3–4, 2003, 299–310 (in Polish).

Dugay A., Herrenknecht C., Czok M., Guyon F., Pages N., New procedure for selective extraction of polycyclic aromatic hydrocarbons in plants for gas chromatographic-mass spectrometric analysis, *Journal of Chromatography A*, 958, 2002, 1–7.

Dymaczewski Z., Jeż-Walkowiak J., eds., *Water supply, quality and waters protection*, Wydawca PZiTS, Poznań, 2012 (in Polish).

Ferrarese E., Andreattola G., Oprea I.A., Remediation of PAH-contaminated sediments by chemical oxidation, *Journal of Hazardous Materials*, 152, 2008, 128–139.

Gdaniec-Pietryka M., Wolska L., Namieśnik J., Physical speciation of polychlorinated biphenyls in the aquatic environment, *Trends in Analytical Chemistry*, 10, 2007, 1005–1012.

Grajek H., Factors influencing on the effectiveness and selectivity of extraction to the solid phase, presented at Conference on the Micropollutants in the Human Environment, Conference 51, Czestochowa University of Technology, Czestochowa, 2003, 347–371 (in Polish).

Grung M., Holth T.F., Jacobsen M.R., Polycyclic aromatic hydrocarbon (PAH) metabolites in Atlantic cod exposed via water or diet to a synthetic produced water, *Journal of Toxicology and Environmental Health A*, 72, 2009, 254–265.

Haritash A.K., Kaushik C.P., Biodegradation aspects of polycyclic aromatic hydrocarbons (PAHs): A review, *Journal of Hazardous Materials*, 169, 2009, 1–15.

Harman Ch., Holth T.F., Hylland K., Thomas K., Grung M., Relationship between polycyclic aromatic hydrocarbon (PAH) accumulation in semipermeable membrane devices and PAH bile metabolite levels in Atlantic cod (*Gadus morhua*), *Journal of Toxicology and Environmental Health A*, 72, 2009, 234–243.

Helaleh M.I.H., Al-Omair A., Nisar A., Gevao B., Validation of various extraction techniques for the quantitative analysis of polycyclic aromatic hydrocarbons in sewage sludges using gas chromatography-ion trap mass spectrometry, *Journal of Chromatography A*, 1083, 2005, 153–160.

Hua L., Wu W.X., Liu Y.X., Tientchen C.M., Chen Y.X., Heavy metals and PAHs in sewage sludge from twelve wastewater treatment plants in Zhejiang Province, *Biomedical and Environmental Science*, 21, 2008, 345–352.

Jamróz T., Ledakowicz S., Miller J., Sencio B., Toxicity of polycyclic aromatic hydrocarbons and the products of their decomposition, *Engineering and Chemical Apparatus* 3s, 2002, 45–46 (in Polish).

Jin W., Toutianoush A., Tieke B., Size- and charge-selective transport of aromatic compounds across polyelectrolyte multilayer membranes, *Applied Surface Science*, 246, 2005, 444–450.

Kalinowski R., Załęska-Radziwiłł M., Assessment of PAH toxicity toward freshwater organisms, presented at Conference on the Micropollutants in the Human Environment, Conference 56, Czestochowa University of Technology, Częstochowa 2005, 48–56 (in Polish).

Kowalczyk D., Świetlik R., Dojlido J.R., Contamination of surface waters with polycyclic aromatic hydrocarbons in Radomka River Basin, presented at Conference on the Micropollutants in the Human Environment with Respect to EU Regulations, Conference 39, Czestochowa University of Technology, Częstochowa 2000, 60–66 (in Polish).

Kułakowski P., Priority substances and other contaminants in water environment, 2009, www.wios.tarnow.pl (in Polish).

Lee H.S., Yoo J.W., Removal of polyaromatic hydrocarbons from scrap tires by solvent extraction, *Korean Journal of Chemical Engineering*, 4, 2011, 1065–1069.

Librando V., Sarpietro M.G., Castelli F., Role of lipophilic medium in the absorption of polycyclic aromatic compounds by biomembranes, *Environmental Toxicology and Pharmacology*, 14, 2003, 25–32.

Ling W., Zeng Y., Gao Y., Dang H., Zhu X., Availability of polycyclic aromatic hydrocarbons in aging soils, *Journal of Soils and Sediments*, 10, 2010, 799–807.

Makhniashvili I., Nitro derivatives of polycyclic aromatic hydrocarbons in the environment, *Work Security*, 3, 2003, 17–20 (in Polish).

Martorell I., Perello G., Marti-Cid R., Castell V., Llobet J.M., Domingo J.L., Polycyclic aromatic hydrocarbons (PAH) in foods and estimated PAH intake by the population of Catalonia, Spain: Temporal trend, *Environment International*, 36, 2010, 424–432.

Mechlińska A., Wolska L., Namieśnik J., Isotope-labeled substances in analysis of persistent organic pollutants in environmental samples, *Trends in Analytical Chemistry*, 29, 2010a, 820–831.

Mechlińska A., Wolska L., Namieśnik J., Comparison of different extraction techniques of polychlorinated biphenyls from sediments samples, *Analytical Letters*, 43, 2010b, 1149–1161.

Namieśnik J., Jamrógiewicz Z., Pilarczyk M., Torres L., *Preparation of environmental samples into analysis, Scientific and Technical Publishing*, Warsaw, 2000 (in Polish).

Nowacka A., Włodarczyk-Makuła M., Changes in the concentration of organic substances (selected indicators) and AOX in water during treatment processes, presented at 3rd International Symposium of Green Chemistry for Environment, Health and Development, Skiathos, Greece, 2012.

Nowacka A., Włodarczyk-Makuła M., Monitoring of PAHs in water during preparation processes, *Polycyclic Aromatic Compounds*, 33, 5, 2013, 430–450.

Nowak R., Wiśniowska E., Włodarczyk-Makuła M., Janosz-Rajczyk M., Leaching of PAHs from fly ashes mechanically activated, *Polish Journal of Environmental Studies*, 18, 2B, 2009, 152–154.

Park J.M., Lee S.B., Kim J.P., Kim M.J., Kwon O.S., Jung D.I., Behaviour of PAHs from sewage sludge incinerations in Korea, *Waste Management*, 29, 2009, 690–695.

Pena T., Casais C., Mejuto C., Cela R., Development of a matrix solid-phase dispersion method for the determination of polycyclic aromatic hydrocarbons in sewage sludge samples, *Analytica Chimica Acta*, 626, 2008, 155–165.

Perez S., Guillamon M., Barcelo D., Quantitative analysis of polycyclic aromatic hydrocarbons in sewage sludge from wastewater treatment plants, *Journal of Chromatography*, 938, 2001, 57–65.

Piekarska K. Modifications of Salmonella test into evaluation of mutagenicity of ashes contaminants of the air, Monograph 52, Scientific Investigations of the Institute of Engineering and Environment Protection, Wrocław University of Technology, Wrocław, 2008 (in Polish).

Popenda A., Włodarczyk-Makuła M., Janosz-Rajczyk M., Polycyclic aromatic hydrocarbons in sewage sludge originating from industrial wastewater treatment, *Archives of Environmental Protection*, 3, 2002, 89–98 (in Polish).

Rejman W., Hazard of Wrocław intake infiltration with PAHs, *Water Management*, 8, 1994, 30–35 (in Polish).

Rosik-Dulewska C., Migration of PAHs from unsealed landfill of wastes into ground waters, *Annals of Environmental Protection*, 9, 2007, 335–344 (in Polish).

Smol M., Włodarczyk-Makuła M., Effectiveness in the removal of polycyclic aromatic hydrocarbons from industrial wastewater by ultrafiltration technique, *Archives of Environmental Protection*, 38, 4, 2012a, 49–58.

Smol M., Włodarczyk-Makuła M., The use of reverse osmosis in the removing of PAHs from municipal landfill leachate, presented at XVII (IV Międzynarodowa) Konferencja Naukowo-Techniczna "Advances in Sustainable Sewage Sludge Management," ASSM, Wydawnictwo Politéchniki Częstochowskiej, Częstochowa, 2012b, 98.

State of rivers based on the results of investigations carried out in the frame on national environmental monitoring in years 2007–2009, Library of Environmental Monitoring, Main Inspector of Environmental Protection, Warszawa, 2010 (in Polish).

Szymański K., Siebielska I., Analytical aspects of the estimation of quality of polluted ground waters, *Environmental Protection*, 1, 2000, 15–18 (in Polish).

Traczewska T.M., Biotoxicity of microbiological products of fates of anthracene and phenenthrene in water and the possibility of their removal, Oficyna Wydawnicza Politechniki Wrocławskiej, Wrocław, 2003.

Turek A., Włodarczyk-Makuła M., The removal of PAHs (C13-C16) from industrial wastewater with the usage of dioxide dihydrogen, Zeszyty Naukowe Uniwersytetu Zielonogórskiego, *Serie Environmental Engineering*, Zielona Góra, 145/25, 2012, 56–64 (in Polish).

Twardowska I., Włodarczyk-Makuła M., Occurrence of leaching of polycyclic aromatic hydrocarbons (PAH) generated at steelworks, *Central European Journal of Public Health*, 1, 2, 1993, 117–120.

Włodarczyk-Makuła M., Quantitive changes of PAHs during wastewater and sludges treatment, Monograph 126, Wydawnictwo Politechniki Częstochowskiej, Częstochowa, 2007 (in Polish).

Włodarczyk-Makuła M., Preparation of wastewater samples for GC analysis of PAHs, *Archives of Environmental Protection*, 3, 2008a, 259–264.

Włodarczyk-Makuła M., PAHs balance in solid and liquid phase of sewage sludge during fermentation process, *Journal of Environmental Science and Health A*, 14, 2008b, 1602–1609.

Włodarczyk-Makuła M., Application of UV rays to removal of polycyclic aromatic hydrocarbons from treated wastewater, *Journal of Environmental Science and Health A*, 46, 3, 2011a, 248–257.

Włodarczyk-Makuła M., Quantitive changes of PAHs in treated wastewater during oxidation, *Archives of Environmental Protection*, 13, cz. 2, 2011b, 1093–1104.

Włodarczyk-Makuła M., Janosz-Rajczyk M., The impact of liming of sewage sludges on the leaching of selected organic micropollutants, *Engineering and Environmental Protection*, 3, 2007, 217–229 (in Polish).

Włodarczyk-Makuła M., Turek A., The application of dioxide dihydrogen in removal of PAHs from coke plants wastewater, monographers of Committee of Environmental Engineering Polish Academy of Science, nr 100, 2012, 349–357 (in Polish).

Wolska L., Determination (monitoring) of PAHs in surface waters: Why operationally defined procedure is needed, *Analytical and Bioanalytical Chemistry*, 391, 2008, 2647–2652.

Wolska L., Gdaniec-Pietryka M., Konieczka P., Namieśnik J., Problems of PAH quantification by GC-MS method using isotope-labelled standards, *Talanta*, 78, 2009, 730–735.

Wolska L., Rawa-Adkonis M., Namieśnik J., Determining PAHs and PCBs in aqueous samples: Finding and evaluating sources of error, *Analytical and Bioanalytical Chemistry*, 382, 2005, 1389–1397.

Xia X., Li G., Yang Z., Chen Y., Huang G.H., Effects of fulvic concentration and origin on photodegradation of polycyclic aromatic hydrocarbons in aqueous solution: Importance of active oxygen, *Environmental Pollution*, 157, 2009, 1352–1359.

Yakan S.D., Karacik B., Ceylan D., Dogu S., Okay O.S., Okay O., Sorption kinetics of PAH by using various sorbents, in *Proceedings of the 2nd International CEMEPE and SECOTOX Conference*, ed. A. Kungolos, K. Aravossis, A. Karagiannidis, P. Samaras, Greece-Mykonos, 2009, 1385–1390.

Załęska-Radziwiłł M., Łebkowska M., Kalinowski R., Biodegradation of selected polycyclic aromatic hydrocarbons, *Environmental Protection and Natural Resources*, Warszawa 2007 (in Polish).

Zhao X., Wang Y., Ye Z., Borthwick A.G.I., Ni J., Oil field wastewater treatment in biological aerated filter by immobilized microorganisms, *Process Biochemistry*, 41, 2006, 1475–1485.

Zhou H.-C., Zhong Z.-P., Jin B.-S., Huang Y.-J., Xiao R., Experimental study on the removal of PAHs using in-duct activated carbon injection, *Chemosphere*, 59, 2005, 861–869.

Directives and Regulations

European Parliament:

  Directive of the European Parliament and Council (*Official Journal WE L* 348, 2008)

  Directive of the European Parliament and Council WE Nr 850/2004 on the persistent organic pollutants, changing the Directive 79/117/EWG (*Official Journal WE L* 158, 2004)

U.S. Environmental Protection Agency, Washington, DC:

  EPA Method 3520C, Continuous liquid-liquid extraction of organics

  EPA Method 3540C, Soxhlet extraction

  EPA Method 3541, Automated Soxhlet extraction

  EPA Method 3545A, Pressurized fluid extraction (PFE)

  EPA Method 3546, Microwave extraction

  EPA Method 3550C, Ultrasonic extraction

  EPA Method 3561, Supercritical fluid extraction of polynuclear aromatic hydrocarbons

  EPA Method 3611B, Alumina column cleanup and separation of petroleum wastes

  EPA Method 3630C, Silica gel cleanup

  EPA Method 3640A, Gel-permeation cleanup

  EPA Method 3650B, Acid-base partition cleanup

  EPA Method 3665A, Sulfuric acid/permanganate cleanup

  EPA Method 8100, Polynuclear aromatic hydrocarbons

  EPA Method 8310, Polynuclear aromatic hydrocarbons (18 compounds)

International Organization for Standardization (ISO), Geneva:

  PN-EN ISO 17993:2005, Water quality—determination of 15 PAHs in water using HPLC with fluorescence detection after liquid-liquid extraction

Minister of the Environment, Poland:

  Dz. U Nr 27 poz. 169, 2009—Regulation of the Minister of Environment regarding the conditions that should be fulfilled in wastewater into waters or soil, as well as substances especially dangerous for water environment (*Journal of Laws* no. 137, item. 984, 2006 and *Journal of Laws*. no. 27, item 169, 2009) (in Polish)

Dz. U Nr 143 poz. 896, 2008—Regulation of the Minister of Environment on the criteria and method of evaluation of the status of underground waters (*Journal of Laws* no. 143, item 896, 2008) (in Polish)

Dz. U Nr 162 poz. 1008, 2008—Regulation of the Minister of Environment on the classification status of surface water bodies (*Journal of laws* no. 162, item 1008, 2008) (in Polish)

Dz. U Nr 165 poz. 1359, 2002—Regulation of the Minister of Environment on the standards of the quality of the soil and standards of the quality of the land (*Journal of Laws* no. 165, item. 1359, 2002) (in Polish)

Dz. U Nr 204 poz. 1728, 2002—Regulation of the Minister of Environment on requirements for surface waters used for the supply into waters intended for consumption (*Journal of Laws* no. 204, item. 1728, 2002) (in Polish)

Dz. U Nr 254 poz. 1528, 2011—Regulation of the Minister of Environment on priority substances for water politics (*Journal of Laws* no. 254, item. 1528, 2011) (in Polish)

Dz. U Nr 257 poz. 1545, 2011—Regulation of the Minister of the Environment on the method of classifying the status of uniform bodies of surface water and environmental quality standards for priority substances (*Journal of Laws* no. 257, item 1545, 2011) (in Polish)

Dz. U Nr 258 poz. 1550, 2011—Regulation of the Minister of the Environment on the form and method and ways of conducting monitoring of uniform bodies of surface and ground waters (*Journal of Laws* no. 258, item 1550, 2011) (in Polish)

Dz. U Nr 298 poz. 1771, 2011—Regulation of the Minister of Environment on the underground wastes landfills (*Journal of Laws* no. 298, item. 1771, 2011) (in Polish)

Minister of Health, Poland:

Dz. U Nr 72 poz. 466, 2010—Regulation of the Minister of Health on the quality of water intended for the human consumption (*Journal of Laws* no. 61, item. 417, 2007; *Journal of Laws* no. 72, item. 466, 2010) (in Polish)

Dz. Urz. WE L33/2006—Explanations to Regulation (EC) No 166/2006 of the European Parliament and of the Council of 18 January 2006. On the establishment of a European Register of Pollutant Release and Transfer and amending Council Directive 91/689/EEC (OJ Office, EU L 33, 4.2.)

EPA Method 610, Polynuclear aromatic hydrocarbons

4

PAHs in Water Resources and Environmental Matrices in Tunisia

Olfa Mahjoub[1] and Imen Haddaoui[2]

[1]*National Research Institute for Rural Engineering, Water, and Forestry (INRGREF), Tunis, Tunisia*

[2]*Higher Institute of Agronomic Sciences of Chatt Meriem, Tunis, Tunisia*

CONTENTS

4.1 Introduction

Polycyclic aromatic hydrocarbons (PAHs) encompass a large number of toxic compounds. They are well known for being ubiquitous xenobiotics endowed with carcinogenic, mutagenic, and genotoxic potential (Busetti et al., 2006).

Concern has been raised about PAHs because the metabolic oxidative activation and the transformation of metabolites of PAHs into genotoxic compounds may damage DNA and initiate carcinogenic processes (Behnisch et al., 2001). Benzo(a)pyrene is one of the most toxic compounds, recognized as being carcinogenic in animals and potentially in humans by the World Health Organization's International Agency for Research on Cancer (IARC). This compound is usually used to express the toxicity of the total PAHs present in a given matrix.

Contamination of water resources by PAHs occurs through various pathways. Water bodies can be exposed to PAHs through atmospheric deposition and direct releases of liquid substances and solid material. Nowadays, research on the occurrence of PAHs in water resources is well developed worldwide, especially in marine environments. From this point of view, particular attention was paid to the Mediterranean from both sides of the basin because of the increasingly dreadful status of the marine environment and coastal areas. Tunisia, with its north and east coasts on the Mediterranean Sea, is embarked with other Mediterranean countries in the protection of both its freshwater resources and the marine environment, and has developed research on the occurrence of hydrocarbons in various matrices. Meanwhile, the occurrence of PAHs in conventional water resources (surface water, groundwater) and nonconventional resources (wastewaters produced in treatment plants) is deemed to be relatively recent.

Scientific research has registered substantial progress beginning circa the year 2000. This was made possible thanks to the accessibility of peer-reviewed literature to the scientific communities, the advancement of scientific knowledge, and the availability of analytical techniques for identification and quantification of the chemical compounds. Research work has covered not only water resources but also several matrices, including sediments, soils, sewage sludge as the ultimate sink of PAHs, and biota. To our knowledge and based on an extensive search of the scientific literature, no studies in this geographic area existed on the contamination of water resources by PAHs before the first decade of the 21st century. The first study dealing with PAHs in sediments from Bizerte lagoon was performed in 2001 (Trabelsi and Driss, 2005).

This chapter offers a thorough overview on the state of the contamination by PAHs of water resources and other exposed matrices, such as soils and sediments, in Tunisia. It is focused on the extent of the pollution and its relevance, in addition to potential environmental implications such as the transfer to soils and biota.

The chapter is organized into the following main sections comprising (1) an insight on the water resources and their potential sources of pollution; (2) the occurrence of PAHs in water resources including surface, ground-, and wastewater; (3) the occurrence of PAH solid matrices such as soils, sediments exposed to contaminated water resources, and sludges; and finally, (4) the presence of PAHs in biota.

4.2 Water Resources and Potential Sources of Pollution in Tunisia

Tunisia is, in the great majority of its territory, arid to semiarid. Based on drinking water demand prospective and the population and socioeconomical growth, the water share per person per year is expected to decrease tremendously by the year 2030, reaching an amount as low as 360 m^3.

Such a decrease results from the disparity between the limited available natural water resources and the increasing demand. Mobilizing 90% of the water resources, estimated at 2732 million cubic meters (MCM), was the approach adopted by the national water strategy for the mobilization of water resources established between 1975 and 2010 for closing the existing gap in resources. In a second stage, the long-term water strategy designed for the Horizon 2030 was oriented toward the management of the water demand estimated at 2760 MCM and saving 30% of the conventional water resources. The strategy was also aimed toward preserving water bodies from pollution and using nonconventional water resources, such as brackish water and reclaimed water (ITES, 2014). In fact, from the qualitative point of view, only 50% of the water resources have salinity less than 1.5 g/L; in the south the salinity level is above 2 g/L for 95% of the resources. The competition between domestic, agricultural, and industrial sectors for access to freshwater resources is one of the motivations of using alternative resources, i.e., reclaimed water and drainage water. With agriculture as the major water-consuming sector (85%), nonconventional resources should be increasingly considered, from qualitative and quantitative points of view, to be fully reused for irrigation by the year 2020. In view of the current situation, the reuse of reclaimed water is expected to be better accepted in the future by the end users provided a substantial improvement is seen in the quality delivered.

In addition to the stresses of the quantitative water scarcity and the qualitative degradation caused by salinity, natural water bodies are continuously being polluted by uncontrolled discharges of untreated or insufficiently treated industrial effluents produced in more than 40% of the units not connected to the sewerage network yet. Indeed, industrial effluents are recognized as the main source of pollution of water resources in Tunisia. They represent 20% of the wastewaters processed in treatment facilities. Discharges of industrial effluents in the sewer system are known to impair the efficiency of the treatment process per se and to reduce the quality of the effluents because of the toxic pollutants. Since 1991, it has been mandated in the legal code that all industries must carry out environmental impact assessment studies. In addition, the regulations require that stationary point industrial effluents with a high polluting load undergo pretreatment prior to discharge in the sewerage network. Considering these sources of pollution, the degradation of the water resources was estimated to cost about 0.6% of the Tunisian GDP. To alleviate the pollution burden linked to the disposal of raw wastewater—either domestic or industrial—an environmental national strategy was set based on several components, including access to sanitation, waste management, and industrial pollution management. Combating pollution will be achieved through the monitoring and control of water quality by several agencies, establishment of groundwater vulnerability maps, and inventorying the water pollution sources, in addition to the implementation of an environmental strategy for the qualitative and quantitative conservation of water resources and ecosystems (Hamza, 2010).

Nowadays, the quality of the water resources in Tunisia is monitored by several governmental institutions operating under the Ministry of the Environment, the Ministry of Health, and the Ministry of Agriculture, plus their related agencies. A water quality monitoring system has been established recently. It was initially tested on conventional waters in dams and streams both used for drinking water supply. This kind of network is dedicated to the determination of salinity and nitrate levels, besides other inorganic and organic parameters, in more than 200 stations at a frequency ranging from at least twice per year (for groundwater) to four times per year (for surface waters). Following the identification of pollution hot spots, a network for monitoring water resources pollution was established and finalized in the year 2014.

4.2.1 Surface Water

Based on rainfall distribution and geographic landscape, surface water resources in Tunisia are concentrated in the north (81%) with an amount of 2700 MCM/year. They encompass water flowing in streams and their tributaries, and stored in reservoirs, dams, and lakes. In 2013, there were reported to exist 33 large dams, 225 hill dams, and 800 hill lakes (ITES, 2014). The quality of surface waters is generally acceptable, especially in the north, with 74% of the resources having salinity below 1.5 g/L (Ben Abdallah, 2003). Apart from salinity, point and non-point pollution sources such as atmospheric deposition, agricultural and urban runoff, discharge from industries, and effluents from treatment plants are having a negative impact on the quality of surface water. Pesticides are presumably the most relevant chemical pollutants that could be detected in water resources; emitted by agricultural activity, they are still insufficiently and inefficiently controlled.

4.2.2 Groundwater

Geographically speaking, at the opposite of the surface water concentrated in the north, groundwater resources are mainly located in the center and the south of the country, where the climate is rather semiarid to desert. An amount of 2180 MCM of groundwater can be potentially mobilized, with 745 MCM from shallow aquifers in the north and the center and 1435 MCM from deep nonrenewable aquifers in the south. From an exploitation view, some aquifers are being overpumped, while others still need to be developed (ITES, 2014).

In Tunisia, the quality of groundwater in shallow aquifers is chiefly threatened by salinity and nitrates: the first is generally caused by seawater intrusion in coastal regions where water is overdrawn from wells to be used for irrigation in agricultural areas, whereas the second is a direct result of the intensive use of fertilizers in agriculture, leading to their leaching and transfer with runoff. Except for these two polluting sources, little is mentioned about pollution in reports and peer-reviewed literature. Indeed, the national network for groundwater quality monitoring only takes into consideration these two parameters for the evaluation of the water quality through the whole year. Other potential sources of pollution are solid wastes produced in treatment plants, after they are disposed of in the environment, e.g., along riverbanks, streams, and abandoned lands. When insufficiently treated, biosolids can release harmful and toxic chemical compounds, thereby contaminating water resources when these chemicals are not degraded in the meantime or adsorbed onto soil particles or organic matter.

4.2.3 Wastewater

Wastewaters represent about 6.3% of the total available water resources in Tunisia, and their amount is expected to reach 12.4% by the year 2030. By then, the volume of reclaimed water will increase to attain 300 million m^3/year.

This volume is projected to be directly reused or stored to meet the agricultural demand when conventional water is lacking (ITES, 2014). In Tunisia, domestic and urban origins account for 80% of wastewaters. The 240 MCM produced in 2013 underwent secondary biological treatment in 110 treatment plants; 80% of them are operating in activated sludge. In terms of reuse, Tunisia has pioneered the Arab and African countries in using reclaimed water since 1965, when citrus trees in the region of Soukra, north of Tunisia, started to be irrigated to safeguard the orchards threatened by seawater intrusion. Nowadays, reclaimed water is used at a rate of 20 to 30%. Irrigation of crops is the main domain and covers fruit trees and fodders, in addition to golf courses and recreational areas. The remaining 70 to 80% of the annually produced effluents is discharged in the receiving environment, either into the waterways or into the sea. Since the quality of the effluents has degraded for various reasons, including industrial development, several water resources have become vulnerable. Despite the low proportion of industrial effluents, the industrial polluting load remains significant, especially when effluents are not pretreated beforehand, or when they do not comply with the national standards of discharge in the public hydraulic domain, or in the sewer system even after secondary treatment. Runoff and storm waters have to be considered additional sources of pollution because they leach heavy metals and hydrocarbons from roads and paved ways that end up in the combined sewer systems.

4.3 PAHs in Water Resources

It is widely established in peer-reviewed literature that PAHs originate from incomplete combustion of organic matter, oil, coal, wood, etc. (Paxéus, 1995; Budzinski, 1997; Blanchard et al., 2001). In addition to their natural origins, it is well known that anthropogenic activities are the main sources of PAHs in the environment. A correlation has been found between the population density and the atmospheric deposition of PAHs (Garban et al., 2002).

In Tunisia, industrial activity has shown great progress during the last decades. With the development of industrial units, an increase of illegal discharge has been observed. Natural environments such as streams and lagoons are frequently used to get rid of treated and untreated effluents. These industries are usually located in coastal cities like Bizerte in the north or Sfax in the center of the country. In those cities already recognized to be heavily populated, official reports gave proof that water resources are becoming highly contaminated due to urban activities, ship traffic, municipal and industrial discharge of effluents, fishing activity and aquaculture, discharge of crude oils, agricultural runoff, etc. (Louati et al., 2001; Trabelsi and Driss, 2005).

4.3.1 PAHs in Surface Water

Surface waters are the first milieu receiving PAHs, because these chemical compounds can enter the environment via the atmosphere. Wet and dry depositions of suspended matter are frequently reported in the literature (Garban et al., 2002; Ollivon et al., 2002). Surface waters are the ultimate reservoir of runoff from flushed roofs and paved surfaces, drainage water, discharge from treatment plants, and others. This makes the contamination of surface waters by PAHs one of the main issues that have been extensively investigated worldwide. Numerous studies have also examined the seasonal and temporal distribution of PAHs and their leaching based on their ability to dissolve in liquid phase and adsorb to suspended matter.

In Tunisia, the extensive review of all the available publications on the contamination of environmental matrices by PAHs leads to the conclusion that surface waters (rivers, streams, or lakes) were not investigated during recent research projects. However, data on the concentrations of PAHs in corresponding *sediments* are available in other publications.

4.3.2 PAHs in Groundwater

Migration of PAHs through the soil column to reach the aquifers may occur after artificial recharge using reclaimed water, through soil contamination by hydrocarbons or leakage from oil industries. PAHs are among the dissolved organics that may contaminate groundwater. Solubility, sorption, and degradation are the main phenomena that govern their persistence in aquifers. Contamination of groundwater by PAHs depends also on the soil type and soil solution transfer. Generally, the lower the number of rings in the molecule and the organic matter content, the more probable is the transfer of PAHs through the soil to reach and contaminate groundwater (Bayard et al., 1998). Soil salinity is an additional factor that can enhance solubility (by increasing the ionic force of the soil solution) and transfer of chemical compounds like PAHs through the soil to deeper layers of the aquifers (Kim and Osako, 2003). Dissolved organic matter in soil solution also plays a key role in the contamination of groundwater by PAHs (Weber et al., 2006). Transfer of PAHs to groundwater takes place mainly by adsorption on dissolved organic matter. Thus, the type of organic matter and the soil content are key parameters.

Due to the paucity of the available knowledge and the data published on PAH concentration levels in surface water and groundwater resources in Tunisia, it is difficult to draw conclusions, other than that this is an environmental concern. This knowledge gap calls for more comprehensive studies and publications on the topic in the local context.

4.3.3 PAHs in Wastewaters

Wastewaters are known to act as vehicles for PAHs (Pham and Proulx, 1997; Busetti et al., 2006), especially the compounds of low molecular weight. Phenanthrene, naphthalene, fluorene, pyrene, and chrysene are commonly detected. Compounds with two to three rings may represent between 60 and 80% of the total PAHs that can be detected (Boström et al., 2002), while compounds of high molecular weight tend to be distributed between liquid and solid phases.

According to peer-reviewed published literature, PAHs are usually washed off from roads and transported with runoff. Car exhausts are indeed important sources of PAHs found in wastewater (Tanaka et al., 1990; Masih and Taneja, 2006). Households contribute to the production of PAHs through various activities such as heating, food processing, cooking, etc. Pyrene and phenanthrene may represent up to 60% of the total load. Rainwater also acts as a vehicle for PAHs, chiefly phenanthrene, fluoranthene, pyrene, and chrysene, which could represent up to 70% of the most known 14 PAHs (Garban et al., 2002). This may also explain the seasonal trend observed for the sum of PAH concentrations exhibiting a higher load in winter than in summer due to the use of coal and heating systems.

Despite the huge records of publications on PAHs worldwide, very little information is available on the occurrence of PAHs in wastewaters in Tunisia. In the irrigated area of Oued Souhil, the analysis of 16 priority PAHs showed that only pyrene was quantified at 32 ng/L in the dissolved phase of the reclaimed water used for irrigation (Mahjoub et al., 2011). This was a quantitative analysis performed to identify compounds that exhibited dioxin-like activity beforehand in tested reclaimed water samples. Based on this result, obtained on the dissolved phase of the reclaimed water, higher concentrations of pyrene are expected to be measured in the whole sample. Other compounds, like phenanthrene and acenaphthene, could have been detected, as was the case in Italy with concentrations of pyrene of 2 to 8 ng/L in the dissolved phase and 26 to 156 ng/L after integration of the particulate phase. Indeed, it was shown that in wastewater, PAH-like compounds are more likely to be partitioned between the dissolved and suspended phases (Dagnino et al., 2010). The application of regular secondary treatment processes has never been effective in targeting the removal of emerging pollutants, like PAHs, from effluents because of the wide range of physicochemical properties. Indeed, the rate of removal of PAHs from raw effluents comprises between 4 and 100% (Jiries et al., 2000). To date, there has been no recent research carried out on the occurrence of PAHs in wastewaters in Tunisia. It was assumed that PAHs are more likely adsorbed on sewage sludge, and consequently not worth investigating in the aqueous phase. Based on the numerous research works that have been performed in European countries, PAHs are quantified in wastewaters prior to analysis in sediments and sewage sludge. In Tunisia, the analysis of PAHs in solid matrices, as will be described further in this chapter, was never linked to the analysis of the compounds in wastewaters. Hence, the following section gives a comprehensive overview of the results obtained on the occurrence of PAHs in solid matrices such as soils, sediments in fresh and marine water, and sewage sludge. Concentrations in the aqueous phase will be given when found.

4.4 PAHs in Environmental Solid Matrices

The presence of PAHs in solid matrices is governed by several factors related to their physicochemical properties. The type of the matrix and its characteristics, including particle size and organic matter content, are deemed as major determinants. Adsorption/desorption processes are governing the accumulation of the molecules or their release to the aqueous phase.

In the following, the levels of contamination of solid matrices by PAHs after exposure to contaminated waters are presented, to showcase how these matrices can reflect on the evolution of the quality of the water resources through time. The concentrations of PAHs that can be measured in solid matrices correspond to an integrative response; therefore, the measured values are more accurate in space than in time.

4.4.1 PAHs in Sediments

Based on their properties, sediments are the sinks and the ultimate reservoir for PAHs. Consequently, they can be considered for monitoring their load to the aquatic environment and evaluating its quality (Prahl et al., 1984; Trabelsi and Driss, 2005).

The presence of PAHs in Tunisian sediments collected from lakes, surface water, and the coastline has started to be investigated only during the last decade. Since marine sediments were particularly concerned, many studies were carried out recently on various sites located along the coastline because of the anthropogenic pressure exerted by tourism, urbanization, fishing, oil industries, and other activities. In the following, examples showcase the extent of sediments' pollution in various regions in Tunisia. It is also worth mentioning that in almost all case studies, the analysis of sediments was not systematically accompanied by analysis of the contaminated water resource regardless of the studied source. This reduces the chances of realistic interpretation that takes into account key factors such as seasonal trend of pollution, spatial and temporal load of pollutants and their distribution, etc.

4.4.1.1 PAHs in Freshwater Sediments

As previously stated, freshwater resources in Tunisia consist of waters in reservoirs, salty lakes, and freshwater lakes and streams.

In the north of Tunisia, Bizerte lagoon was the most studied in the last 5 years. Several research teams have focused on this ecosystem because it is classified as a very sensitive environment, prone to multiple sources of pollution. The lagoon is a lake connected to the Mediterranean Sea and to the protected area of Ichkeul Lake. With an area of about 15 km², the lake offers a great variety of activities, like fishing and aquaculture (Ben Ameur et al., 2012). The lake is undergoing high anthropogenic pressure and receiving effluents from wastewater treatment plants, agricultural areas, industrial areas, and urban centers. The temporal and spatial monitoring of the 16 PAHs regulated by the USEPA investigated in 18 stations located around the lake showed that the total concentrations ranged between 83 and 447 ng/g dw, with a mean of 218 ng/g dw (Table 4.1). Phenanthrene, acenaphthene, and anthracene were found in all sediments. A high contamination corresponding to the maximum registered values was deemed to be linked to industrial activities of metal, naval construction, and tire production in Menzel Bourguiba city (Trabelsi and Driss, 2005). Using the ratios of phenanthrene/anthracene and fluoranthene/pyrene as recommended by previous research (Budzinski, 1997) showed that PAHs of pyrolitic origin were prevailing.

TABLE 4.1

Concentrations of PAHs in Sediments of Surface Water

Location	Site	No. of PAHs	Concentration	Reference	Level of Contamination
Bizerte	Lagoon	16 PAHs	83–447 ng/g	Trabelsi and Driss (2005)	Low–moderate
		16 PAHs	2–537 ng/g	Louiz et al. (2008)	Low–moderate
		20 PAHs	20–449 ng/g	Trabelsi et al. (2012)	Not contaminated
Monastir	Oued Leya		40–65 mg/kg; 0–56 mg/kg	Dridi et al. (2014)	—
Sfax	Sebkat Boujmel (salty lake)		11–1124 mg/kg	Aloulou et al. (2011)	—

Contamination with oil was detected in only one site, and was thus judged not relevant. Urbanization was also identified as a source of contamination (Trabelsi and Driss, 2005). A comparison with sediments collected and analyzed elsewhere revealed that the contamination evaluated at that time was low to moderate.

From the qualitative point of view of the pollution load in Bizerte lagoon, the sediments were found to exhibit high estrogenic and antiandrogenic activities in a well-known contaminated site, i.e., Menzel Bourguiba, in addition to dioxin-like activity. Dioxin-like activity is induced by coplanar and hydrophobic compounds like some PAHs (Hilscherova et al., 2000) and are usually worth investigation in solid matrices like sediments (Michallet-Ferrier et al., 2004). The low 7-ethoxyresorufin-O-deethylase (EROD) activity observed on extracts after 24 h exposure, compared to 4 h, indicates the presence of PAH-like compounds that are rapidly biotransformed. Nevertheless, the quantitative analysis showed that the sum of 16 priority PAHs was low to moderate, ranging between 2 and 537 ng/g dw, and contributing only by 4% to the dioxin-like activity. In fact, PAH metabolites may also induce estrogenic activity (Louiz et al., 2008). It should be noted that using in vitro bioassays for evaluating the toxicological pathway of PAHs is cell line dependent (Brasseur et al., 2008), although benzo(a)pyrene is usually used as a PAH reference compound.

A comprehensive study on Bizerte lagoon was performed on 15 surface sediment samples taken on the coast and across the lagoon. The results of the analysis of 20 PAHs, including three alkyl-substituted homologues, showed that concentrations ranged between 20 and 449 ng/g dw. The lowest values were found in the central part of the lagoon, in areas connected to small wadis, and close to agricultural areas. Urban and industrial activities were at the origin of the high concentrations observed in the other stations. Based on ratios between the different congeners allowing the classification of contamination source, the petrogenic and pyrogenic sources were found to be dominant. When comparing the observed values to the standards set by the U.S. National Oceanic and Atmospheric Administration, it appeared that Bizerte lagoon is not contaminated because the total concentration is below the low effects range corresponding to 4000 ng/g (Trabelsi et al., 2012). To confirm partly the results described above on Bizerte lagoon, it is worth mentioning that recently, PAHs were detected in air samples taken from the city with a total concentration between 9.4 and 44.8 ng/m^3. A clear abundance of pyrene, fluoranthene, benzo(g,h,i)perylene, benzo(b)fluoranthene, chrysene, and benzo(a)pyrene was noticed (Ben Hassine et al., 2014). Therefore, wet depositions should be considered as a non-point source of contamination of surface waters (Zhu et al., 2004), and consequently sediments.

The contamination of sediments in streams and rivers is less studied in Tunisia. In surface water phenanthrene, anthracene, fluoranthene, pyrene, benzo(a)anthracene, perylene, benzo(e)pyrene, benzo(a)pyrene, and chrysene were detected in Laya stream (Sousse, centre of Tunisia) at a total concentration of 40.7 mg/kg dw. The direct discharges of wastewater, either treated or raw, were identified as the main sources of pollution. Besides, the stream is subjected to direct dumping of sewage sludge from the wastewater treatment plant nearby. Analysis of sediments collected downstream gave a concentration of 0.56 mg/kg dw (Dridi et al., 2014).

In the south of the country, the salty lake (Sebkhat) Boujemal located in the Sfax region was found to be the discharge point of a French-Tunisian oil company situated in the vicinity, operating since 1971. In this closed system, with no connection to the sea, the results have shown the presence of PAHs in surface sediments. Superficial sandy sediments contained 11 to 1124 mg/kg dw of PAHs, with the highest levels registered close to the discharge point of the oil company. The ratios of phenanthrene/anthracene, fluoranthene/pyrene, and Σ methyl-phenanthrenes/phenanthrene have given evidence that the contamination is of petrogenic and pyrogenic origin (Aloulou et al., 2011).

4.4.1.2 PAHs in Marine Sediments

Because the quality of a marine environment is inextricably linked to economic activities such as fishing and tourism, and because the seashore is the ultimate discharge point of all water bodies, marine sediments are the most exposed to urban pollution; for this reason, they are more studied in Tunisia. In the frame of the Programme for the Assessment and Control of Marine Pollution in the Mediterranean (MED POL) launched in 1981, monitoring was performed at a number of sites, dispersed across the country. Contamination of marine sediments with PAHs has become a concern. In order to remedy the situation, a set of standards for marine sediment quality was proposed in 2009 with two reference levels for total PAHs equivalent to 1.7 and 9.6 mg/kg of total PAH/kg of sediment, respectively. Fluoranthene, benzo(b)fluoranthene, benzo(k)fluoranthene, benzo(a)pyrene, benzo(g,h,i)perylene, and indeno(1,2,3-c,d)pyrene were the six PAHs

TABLE 4.2

Concentrations of PAHs in Sediments of Marine Environment

Location	Site	No. of PAHs	Concentration	Reference
Monastir	5 sites in Monastir Bay	17 PAHs	577 ng/g	Nouira et al. (2013a)
Teboulba	Fishing harbor	24 PAHs	16,000 µg/g	Jebali et al. (2014)
Khniss	Coastline	17 PAHs	240–680 µg/g	Zrafi et al. (2013)
Sfax-Kerkenah	15 sites on the coast	17 PAHs	113–1072 ng/g	Zaghden et al. (2007)

considered. Monitoring of these compounds in streams and lagoons showed a decrease between 2007 and 2013. In Bizerte lagoon, for example, values dropped from 0.583 to 0.00772 µg/g (European Environment Agency, 2014).

Table 4.2 shows some concentrations of PAHs in marine sediments that are reported in the literature.

In central Tunisia, some coastal areas were studied because of the anthropogenic pressure. Contamination of sediments from Monastir Bay was investigated because of the high organic and mineral pollution (Nouira et al., 2013a; Souissi et al., 2014). The analysis of 17 PAHs in samples of surface sediments collected from five sites has shown that they contain up to 577 ng/g dw. However, no ecotoxicological effect was expected given that the concentrations are within the sediment quality guidelines. The origins of PAHs were found to be mainly from pyrolitic sources. As concluded, the trend observed in concentrations was related to the hydrological and morphological conditions of the sites (Nouira et al., 2013b). Recently, some initiatives have been taken for the depollution of the Bay of Monastir. The guide used is that published in the 1990s by the USEPA to manage contaminated sediment (USEPA, Sediment Oversight Technical Committee, 1990).

In Teboulba fishing harbor, located downstream from Monastir, it was found that sediments were contaminated with PAHs with concentrations reaching 16,000 µg/g. Sites located downstream and in the middle section of the harbor contained 15 of the 24 PAHs investigated. These were the 16 priority PAHs with additional compounds, and a number of metabolites. Among those detected, dibenzo(a,h)anthracene, benzo(a)pyrene, chrysene, pyrene, anthracene, and phenanthrene were found at high concentrations, which may negatively affect the ecological system. Based on sediment quality guidelines and on the effect range median (ERM) used to assess toxicity of PAHs, those compounds could be harmful because the ERMs were very low compared to their concentrations (Jebali et al., 2014).

In the vicinity of Monastir city, the marine sediments of the Khniss coastline, when collected and analyzed for 17 PAHs, showed a total concentration of aromatic hydrocarbons ranging between 240 and 680 µg/g dw, with a concentration of PAHs in sediments from 6.95 to 14.59 ng/g dw. These values were relatively high compared to those observed in the Mediterranean coast with a mixed contamination of petrogenic and pyrolytic origin. Compounds like fluorene, pyrene, 1-methyl-pyrene, perylene, and chrysene were abundant. A general domination of four-ring and five-ring PAHs was observed. Pollution was attributed to several sources, such as wastewater and wadi Khniss discharges, port activities, fishing industries, atmospheric emission, and extensive maritime traffic (Zrafi et al., 2013).

Farther toward the south of Tunisia, the first study carried out on PAHs was on the coastal area of Sfax—Kerkennah Island. The detected concentrations were in the range of 113 to 10,720 ng/g dw. The high concentrations were more likely linked to the proximity of the harbor and sewage outfall of the city of Sfax. Pyrolytic and petrogenic PAHs were found in many sediments, with a manifest predominance of PAHs of petrogenic origin, especially close to the harbor of Sfax (Zaghden et al., 2007).

4.4.2 PAHs in Soils

Soils can be contaminated by PAHs through various pathways: atmospheric deposition, industrial emissions, irrigation with wastewater, spread of sewage sludge used as fertilizer, etc. PAHs are not rapidly eliminated from soils, making some of them prone to accumulation. Phenomena such as adsorption on organic matter and interaction with soil particles intervene to make them less available to soil microorganisms, thus reducing their biodegradation (Schwarzenbach and Westall, 1981; Lemière et al., 2001). Thus, five-ringed PAHs are the most likely to be degraded (Johnsen et al., 2005).

In Tunisia, there is limited awareness of the potential negative impacts—in terms of either concentration in the matrix or toxicity to fauna or flora—of PAHs in soils. Despite the clear identification of PAH sources in the environment evidenced by the amount of research work carried out during the last decades, it seems that soil is still not regarded as a potential sink for PAHs, nor as an interface for the contamination of cultivated crops and groundwater underneath.

Based on their physicochemical properties, PAHs are more likely to be adsorbed on solid particles of soil and suspended matter in wastewater (Dagnino et al., 2010). Consequently, wastewater spreading during irrigation represents an additional source of PAHs to soil. To our knowledge, the first research work in Tunisia that evidenced the presence of PAHs in agricultural soils was carried out in the area of Oued Souhil, Nabeul, located in eastern Tunisia, in an area irrigated with reclaimed water for more than 30 years. In this soil, the detection of dioxin-like activity linked to the activation of the aryl hydrocarbon receptor indicated the presence of PAHs or PAH-like compounds. Concentrations of pyrene in soil samples collected in the irrigation channels and under trees were approximately 2 to 3 µg/kg dw. In the same area, reclaimed water is partially used in winter for artificial aquifer recharge. Similar concentrations, around 4 µg/kg dw, were obtained in soil collected from recharge basins where reclaimed water is allowed to infiltrate.

The soil properties were almost the same, rather sandy with poor organic matter content. The analysis of reclaimed water used to infiltrate contained around 32 ng/L of pyrene (Mahjoub, 2009; Mahjoub et al., 2011).

Later, more results were obtained during a study carried out on the area of Zaouiet Sousse, which has been irrigated by reclaimed water since 1989. In the irrigated land, the concentrations of PAHs varied between 46.2 and 129.5 ng/g dw. It was found that the longer the irrigation period, the lower the PAH concentration, because of the degradation and potential migration through the soil. In comparison with the control soil, lower amounts were registered with 66.2 ng/g dw for 16 PAH concentrations (Khadhar et al., 2013).

More research studies need to be conducted to investigate PAHs in various soils either in cultivated land or located in potentially contaminated sites such as landfills.

4.4.3 PAHs in Sewage Sludge

The focus for the study of PAH distribution in Tunisian sewage sludge is on their toxicity and persistence in soil after the use of biosolids as organic fertilizers, rather than on their negative impact after discharge to the environment. Biosolids used to be spread on farmland in Tunisia and then stopped in 1998. When it restarted during the 2000s, very little research was found on organic pollutants like PAHs. While the Tunisian national standards took up the subject in 2002 (NT 106.020, 2002), the concentration of PAHs was not subjected yet to threshold values. One of the first studies carried out at the Charguia treatment plant showed that samples of sewage sludge were free of PAHs (El Hammadi et al., 2007). Later, in the context of reuse of sewage sludge in agriculture, the characterization of nine sewage sludges from different treatment plants was fully performed for the first time. The results showed that the sum of PAHs ranged from 95.6 to 7718 ng/g dw (Khadhar et al., 2010). According to the little information available on PAHs in sewage sludge, soil contamination caused by their use as fertilizers has not been examined yet.

Either present in contaminated soils or in aquatic environments, living organisms can be affected by PAHs.

4.5 PAHs in Biota

Contamination of sediments with PAHs has a negative effect on biota. Biomass can be significantly reduced, and some species may completely disappear from the ecosystem. This is the case of some polychaetes and copepods; nematodes were found to be more resistant even though they were affected (Louati et al., 2014).

In Tunisia, the impact of PAH-contaminated water resources on aquatic organisms was studied using several tools. Biomarkers applied at different levels—molecular, biochemical, physiological, and individual—have been increasingly used for the last 5 years, offering an easy, relatively affordable, and reliable technique to identify stressors in the environment and the responses of exposed organisms, as well as burden to biota. In almost all studies, except those carried out in controlled conditions, PAHs were analyzed in biota, but not in the contaminated waters in which the organisms are living, to which they are exposed or which they are consuming.

The best-known case study was carried out in Bizerte lagoon. A genotoxic effect was detected in several fish species through analysis of their DNA. The contamination was found not to be directly linked to PAH contamination, but rather due to a mixture of stressors that could encompass PAHs (Mahmoud et al., 2010). PAHs, among other organic and inorganic compounds, may induce the production of reactive oxygen species (Livingstone, 2001). Bioaccumulation of PAHs occurs in various organs, especially mussels and glands such as digestive glands. The latter are usually used to monitor the metabolism and biotransformation of organic compounds. In a laboratory experiment carried out on caged animals exposed to 19 mg/L per animal, considered to be a sublethal dose, benzo(a)pyrene was found to induce DNA damage after 72 h of exposure (Banni et al., 2010).

Other researchers have studied the effect of PAHs on enzymatic activities (phase I biotransformation) and found a positive correlation between PAH content in the digestive glands of mussels and the detoxification activity phase I benzo(a)pyrene hydroxylase ($r = 0.58$), and a negative correlation with phase II glutathione S-transferase ($r = -0.84$). The total concentration of the 14 PAHs was around 2 μg/g dw (Kamel et al., 2014).

In an attempt to combine chemical and biochemical tools for the assessment of the impact of PAHs on wildlife in the area of Bizerte, 24 compounds were analyzed in fish mussels. Total PAH concentrations varied between 47.5 ± 3.2 ng/g dw and 390.0 ± 57.3 ng/g dw. PAHs with two to three and four to five rings were clearly bioaccumulated through either food ingestion or contact. The most contaminated site was located close to the mouth of a stream crossing an agricultural area, on the route of commercial cargo vessels traveling between two important cities: Bizerte and Menzel Bourguiba (Barhoumi et al., 2014).

The assessment of biota contamination by PAHs in a number of countries of the Mediterranean Basin, including Tunisia, has led to a number of conclusions about the state of contamination of the Mediterranean Basin. In Tunisia, it was found that after 12 weeks of exposure to PAHs, the concentrations of the 16 studied compounds in mussels sampled along the coast ranged between 22 and 106 μg/kg dw in the 123 monitored stations. The two stations located in the Bay of Tunis registered concentrations of 68.6 and 69.7 μg/kg dw. These values were considered among the highest encountered in the Mediterranean Basin. Benzo(a)pyrene and naphtalene were also detected and linked to industrial activity. Levels of chrysene, fluoranthene, and pyrene were high in caged and native mussels due to a probable contamination by steel smelter and coke oven industrial activity. The levels of chrysene were also high in the Bay of Tunis (9.7 and 7.3 μg/kg), with comparable values to those found in Ibiza, Spain (7.1 μg/kg) and Naples, Italy (7.3 μg/kg). For fluoranthene, 11.0 and 10.1 μg/kg were detected, and were deemed among the highest values ever met. Phenanthrene was also high, at 10.3 and 8.8 μg/kg; pyrene was also found, at 9.5 μg/kg. Based on these values, Tunis Bay was considered to be a hot spot of pollution (Galgani et al., 2011).

Despite the manifest high values, biomonitoring of PAHs is still not systematically implemented in the marine environment. It is localized and restricted to limited actions.

4.6 Conclusion

PAHs are ubiquitous pollutants unevenly distributed in the environment. Water resources are prone to contamination by PAHs mainly through discharge of wastes and wastewaters in the receiving bodies. Soils and sediments in freshwater and marine environment can be exposed to PAHs through contaminated water resources and also through dry and wet depositions.

In Tunisia, PAHs were studied, but not regulated or actively combated, despite the worldwide information existing on their toxicity to the environment and the ecological system as a whole. The state of the art presented in this chapter on the contamination of water resources and exposed matrices by PAHs leads to the conclusion that the marine environment is the most affected by PAHs with high contamination levels. It has received much more attention. The contamination of water resources is still not receiving the attention it requires, despite the increasing number of polluting sources such as industries.

Since research is still not well developed and only scattered data are available, regulations are not well established, and the existing ones are not well enforced for a better management and protection of water resources and the exposed matrices.

References

Aloulou F., Kallel M., Belayouni H. (2011). Impact of oil field-produced water discharges on sediments: A case study of Sebkhat Boujemal, Sfax, Tunisia. *Environmental Forensics* 12: 290–299.

Banni M., Negri A., Dagnino A., Jebali J., Ameur S., Boussetta H. (2010). Acute effects of benzo(a)pyrene on digestive gland enzymatic biomarkers and DNA damage on mussel *Mytilus galloprovincialis*. *Ecotoxicology and Environmental Safety* 73: 842–848.

Barhoumi B., Menach K.L., Clérandeau C., Ben Ameur W., Budzinski H., Driss M.R., Cachot J. (2014). Assessment of pollution in the Bizerte lagoon (Tunisia) by the combined use of chemical and biochemical markers in mussels, *Mytilus galloprovincialis*. *Marine Pollution Bulletin* 84: 379–390.

Bayard R., Barna L., Mahjoub B., Gourdon R. (1998). Investigation of naphtalene sorption in soils fractions using batch and columns assays. *Environment and Toxicological Chemistry* 17: 2383–2390.

Behnisch P.A., Hosoe K., Sakai S.-I. (2001). Combinatorial bio/chemical analysis of dioxin and dioxin-like compounds in waste recycling, feed/food, humans/wildlife and the environment. *Environment International* 27: 495–519.

Ben Abdallah S. (2003). La réutilisation des eaux usées traitées en Tunisie. 1ère Partie. Point de départ, conditions-cadres et stratégie politique d'eau. Institut Allemand de Developpement (IAD)/Centre International des Technologies de l'Environnement de Tunis (CITET), Bonn.

Ben Ameur W., de Lapuente J., El Megdiche Y., Barhoumi B., Trabelsi S., Lydia Camps L., Serret J., Ramos-Lopez D., Gonzalez-Linares J., Driss M.R., Borras M. (2012). Oxidative stress, genotoxicity and histopathology biomarker responses in mullet (*Mugil cephalus*) and sea bass (*Dicentrarchus labrax*) liver from Bizerte lagoon (Tunisia). *Marine Pollution Bulletin* 64: 241–251.

Ben Hassine S., Hammami B., Ben Ameur W., El Megdiche Y., Barhoumi B., Driss M.R. (2014). Particulate polycyclic aromatic hydrocarbons (PAH) in the atmosphere of Bizerte City, Tunisia. *Bulletin of Environmental Contamination and Toxicology* 93(3): 375–382.

Blanchard M., Teil M.-J., Ollivon D., Garban B., Chesterikoff C., Chevreuil M. (2001). Origin and distribution of polyaromatic hydrocarbons and polychlorobiphenyls in urban effluents to wastewater treatment plants of the Paris area (France). *Water Research* 35: 3679–3687.

Boström C.-E., Gerde P., Hanberg A., Jernström B., Johansson C., Kyrklund T., Rannug A., Tornqvist M., Victorin K., Westerholm R. (2002). Cancer risk assessment, indicators, and guidelines for polycyclic aromatic hydrocarbons in the ambient air. *Environmental Health Perspectives* 110: 451–489.

Brasseur C., Melens D., Muller M., Maghuin-Rogister G., Scippo M.-L. (2008). Study of the 16 EU-JECFA PAHs interaction with the aryl hydrocarbon receptor (AhR) using rat and human reporter cell lines. *Toxicology Letters* 172: S37–S38.

Budzinski H. (1997). Cycles biogéochimiques des composés aromatiques. Lettre des sciences chimiques. No. 60.

Busetti F., Heitz A., Cuomo M., Badoer S., Traverso P. (2006). Determination of sixteen polycyclic aromatic hydrocarbons in aqueous and solid samples from an Italian wastewater treatment plant. *Journal of Chromatography A* 1102: 104–115.

Dagnino S., Gomez E., Picot B., Cavaillès V., Casellas C., Balaguer P., Fenet H. (2010). Estrogenic and AhR activities in dissolved phase and suspended solids from wastewater treatment plants. *Science of the Total Environment* 408: 2608–2615.

Dridi L., Majdoub R., Ghorbel F., M.B.H. (2014). Characterization of water and sediment quality of Oued Laya (Sousse/Tunisia). *Journal of Material and Environmental Sciences* 5(5): 1500–1504.

El Hammadi M.A., Trabelsi M., Jrad A., Hanchi B. (2007). Analysis and comparison of toxicants between Tunisian activated sludge and produced compost. *Research Journal of Environmental Sciences* 1: 122–126.

European Environmental Agency (2014). Horizon 2020 Mediterranean Report. Annex 6: Tunisia. EEA Technical Report No. 6/2014. 28. Luxembourg: Publications Office of the European Union.

Galgani F., Martinez-Gomez C., Giovanardi F., Romanelli G., Caixach J., Cento A., Scarpato A., Ben Brahim S., Messaoudi S., Deudero S., Boulahdid M., Benedicto J., Andral B. (2011). Assessment of polycyclic aromatic hydrocarbon concentrations in mussels (*Mytilus galloprovincialis*) from the western basin of the Mediterranean Sea. *Environmental Monitoring and Assessment* 172: 301–317.

Garban B., Blanchoud H., Motelay-Massei A., Chevreuil M., Ollivon D. (2002). Atmospheric bulk deposition of PAHs onto France: Trends from urban to remote sites. *Atmospheric Environment* 36: 5395–5403.

Hamza M. (2010). L'eau en Tunisie: Expériences et Priorités. www.medaquaministe-rial2010.net/.../downl... (accessed October 9, 2010).

Hilscherova K., Machala M., Kannan K., Blankenship A.L., Giesy J.P. (2000). Cell bioassays for detection of aryl hydrocarbon (AhR) and estrogen receptor (ER) mediated activity in environmental samples. *Environmental Science and Pollution Research* 7: 159–171.

ITES. (2014). Etude stratégique: Système hydraulique de la Tunisie à l'horion 2030. ITES, Tunis.

Jebali J., Chicano-Galvez E., Fernandez-Cisnal R., Banni M., Chouba L., Boussetta H., Lopez-Barea J., Alhama J. (2014). Proteomic analysis in caged Mediterranean crab (*Carcinus maenas*) and chemical contaminant exposure in Téboulba Harbour, Tunisia. *Ecotoxicology and Environmental Safety* 100: 15–26.

Jiries A., Hussain H., Lintelmann J. (2000). Determination of polycyclic aromatic hydrocarbons in wastewater, sediments, sludge and plants in Karak province, Jordan. *Water, Air, and Soil Pollution* 217–228.

Johnsen A.R., Wick L.Y., Harms H. (2005). Principles of microbial PAH-degradation in soil. *Environmental Pollution* 133: 71–84.

Kamel N., Burgeot T., Banni M., Chalghaf M., Devin S., Minier C., Boussetta H. (2014). Effects of increasing temperatures on biomarker responses and accumulation of hazardous substances in rope mussels (*Mytilus galloprovincialis*) from Bizerte lagoon. *Environmental Science and Pollution Research* 21: 6108–6123.

Khadhar S., Charef A., Hidri Y., Higashi T. (2013). The effect of long-term soil irrigation by wastewater on organic matter, polycyclic aromatic hydrocarbons, and heavy metals evolution: Case study of Zaouit Sousse (Tunisia). *Arabian Journal of Geosciences* 6: 4337–4346.

Khadhar S., Higashi T., Hamdi H., Matsuyama S., Charef A. (2010). Distribution of 16 EPA-priority polycyclic aromatic hydrocarbons (PAHs) in sludges collected from nine Tunisian wastewater treatment plants. *Journal of Hazardous Materials* 183: 98102.

Kim Y.-J., Osako M. (2003). Leaching characteristics of polycyclic aromatic hydrocarbons (PAHs) from spiked sandy soil. *Chemosphere* 51: 387–395.

Lemière B., Seguin J.J., Guern C.L., Guyonnet D., Baranger P., Darmendrail D., Conil P. (2001). Guide sur le comportement des polluants dans les sols et des nappes. Application dans un contexte d'Evaluation Détaillée de Risques pour les ressources en eau. BRGM, Orléans, France.

Livingstone D.R. (2001). Contaminant-stimulated reactive oxygen species production and oxidative damage in aquatic organisms. *Marine Pollution Bulletin* 42: 656–666.

Louati A., Elleuch B., Kallel M., Saliot A., Dagaut J., Oudot J. (2001). Hydrocarbon contamination of coastal sediments from the Sfax area (Tunisia), Mediterranean Sea. *Marine Pollution Bulletin* 42: 445–452.

Louati H., Ben Said O., Soltani A., Got P., Cravo-Laureau C., Duran R., Aissa P., Pringault O., Mahmoudi E. (2014). Biostimulation as an attractive technique to reduce phenanthrene toxicity for meiofauna and bacteria in lagoon sediment. *Environmental Science and Pollution Research* 21: 3670–3679.

Louiz I., Kinani S., Gouz M.-E., Ben-Attia M., Menif D., Bouchonnet S., Porcher J.-M., Ben-Hassine O.K., Aït-Aïssa S. (2008). Monitoring of dioxin-like, estrogenic and anti-androgenic activities in sediments of the Bizerta lagoon (Tunisia) by means of in vitro cell-based bioassays: Contribution of low concentrations of polynuclear aromatic hydrocarbons (PAHs). *Science of the Total Environment* 402: 318–329.

Mahjoub O. (2009). Characterization of organic micropollutants in treated wastewater and transfer diagnosis to soil and groundwater. Case study of Oued Souhil Region, Nabeul, Tunisia (in french). PhD thesis, Université Montpellier I, Montpellier, France.

Mahjoub O., Escande A., Rosain D., Casellas C., Gomez E., Fenet H. (2011). Estrogen-like and dioxin-like organic contaminants in reclaimed wastewater: Transfer to irrigated soil and groundwater. *Water Science and Technology* 63: 1657–1662.

Mahmoud N., Dellali M., El Bour M., Aissa P., Mahmoudi E. (2010). The use of *Fulvia fragilis* (Mollusca: Cardiidae) in the biomonitoring of Bizerta lagoon: A multi-markers approach. *Ecological Indicators* 10: 696–702.

Masih A., Taneja A. (2006). Polycyclic aromatic hydrocarbons (PAHs) concentrations and related carcinogenic potencies in soil at a semi-arid region of India. *Chemosphere* 65: 449–456.

Michallet-Ferrier P., Aït-Aïssa S., Balaguer P., Dominik J., Haffner G.D., Pardos M. (2004). Assessment of estrogen receptor (ER) and aryl hydrocarbon receptor (AhR) mediated activities in organic sediment extracts of the Detroit River, using in vitro bioassays based on human MELN and teleost PLHC-1 cell lines. *Journal of Great Lakes Research* 30: 82–92.

Nouira T., Risso C., Lassaad, C., Budzinski H., Boussetta H. (2013b). Polychlorinated biphenyls (PCBs) and polybrominated diphenyl ethers (PBDEs) in surface sediments from Monastir Bay (Tunisia, Central Mediterranean): Occurrence, distribution and seasonal variations. *Chemosphere* 93: 487–493.

Nouira T., Tagorti M.A., Budzinski H., Etchebert H., Boussetta H. (2013a). Polycyclic aromatic hydrocarbons (PAHs) in surface sediments of Monastir Bay (Tunisia, Central Mediterranean): Distribution, origin and seasonal variations. *International Journal of Environmental Analytical Chemistry* 1–14.

Ollivon D., Blanchoud H., Motelay-Massei A., Garban B. (2002). Atmospheric deposition of PAHs to an urban site, Paris, France. *Atmospheric Environment* 36: 2891–2900.

Paxéus N. (1995). Organic pollutants in the effluents of large wastewater treatment plants in Sweden. *Water Research* 30: 1115–1122.

Pham T.-T., Proulx S. (1997). PCBs and PAHs in the Montreal Urban Community (Quebec, Canada) wastewater treatment plant and in the effluent plume in the St. Lawrence River. *Water Research* 31: 1887–1896.

Prahl F.G., Crecellus E., Carpenter R. (1984). Polycyclic aromatic hydrocarbons in Washington coastal sediments: An evaluation of atmospheric and riverine routes of introduction. *Environmental Science and Technology* 18: 687–693.

Schwarzenbach R.P., Westall J. (1981). Transport of nonpolar organic compounds from surface water to groundwater. Laboratory sorption studies. *Environmental Science and Technology* 15: 1360–1367.

Souissi R., Turki I., Souissfi F. (2014). Effect of submarine morphology on environment quality: Case of Monastir Bay (Eastern Tunisia). *Carpathian Journal of Earth and Environmental Sciences* 9.

Tanaka H., Onda T.,Ogura N. (1990). Determination of polyaromatic hydrocarbons in urban street dusts and their source materials by capillary gas chromatography. *Environmental Science and Technology* 24: 1179–1186.

Trabelsi Souad, Ben Ameur Walid, Derouiche Abdekader, Cheikh Mohamed and Driss Mohamed Ridha (2012). POP and PAH in Bizerte Lagoon, Tunisia. *Applications of Gas Chromatography*, Dr. REZA Davarnejad (Ed.). Available from http://www.intechopen.com/

Trabelsi S., Driss M.R. (2005). Polycyclic aromatic hydrocarbons in superficial coastal sediments from Bizerte lagoon, Tunisia. *Marine Pollution Bulletin* 50: 344–348.

US EPA, Sediment Oversight Technical Committee (1990). Managing Contaminated Sediments. EPA 506/6–90/002. U.S. Environmental Protection Agency, Washington D.C.

Weber S., Khan S., Hollender J. (2006). Human risk assessment of organic contaminants in reclaimed wastewater used for irrigation. *Desalination* 187: 53–64.

Zaghden H., Kallel M., Elleuch B., Oudot J., Saliot A. (2007). Sources and distribution of aliphatic and polyaromatic hydrocarbons in sediments of Sfax, Tunisia, Mediterranean Sea. *Marine Chemistry* 105: 70–89.

Zhu L., Chen B., Wang J., Shen H. (2004). Pollution survey of polycyclic aromatic hydrocarbons in surface water of Hangzhou, China. *Chemosphere* 56: 1085–1095.

Zrafi I., Bakhrouf A., Rouabhia M., Saidane-Mosbahi D. (2013). Aliphatic and aromatic biomarkers for petroleum hydrocarbon monitoring in Khniss Tunisian-Coast (Mediterranean Sea). *Procedia Environmental Sciences* 18: 211–220.

5

Occurrence, Removal, and Fate of PAHs and VOCs in Municipal Wastewater Treatment Plants: A Literature Review

Aleksandra Jelic, Evina Katsou, Simos Malamis, David Bolzonella, and Francesco Fatone

Department of Biotechnology, University of Verona, Verona, Italy

CONTENTS

Chemical compounds in municipal wastewater consist of a wide range of naturally occurring and man-made organic and inorganic constituents. They include industrial and household chemicals, compounds excreted by humans, and by-products formed during wastewater treatment processes. Certain classes of chemicals have attracted increased attention from the general public and the scientific community for their potential impact on the human health and the environment. These include, among others, polycyclic aromatic hydrocarbons (PAHs) and volatile organic compounds (VOCs).

PAHs are a large class of persistent semivolatile organic pollutants with two or more fused aromatic rings. Certain individual PAHs have been classified by the International Agency for Research on Cancer (IARC) as carcinogenic to animals and probably carcinogenic to humans (Boström et al., 2002). PAHs are present in coal, oil, and tar, and formed mainly as a result of the incomplete combustion of organic matter during industrial and other anthropogenic activities, such as coal and petroleum processing, waste incineration, heating, food preparation, motor vehicle traffic, tobacco smoking, etc. A fraction of PAHs reaches the atmosphere through natural processes such as wood fires and volcano eruptions. PAHs may reach water bodies via atmospheric deposition, rainwater runoff, and municipal and industrial wastewater effluents, as well as via oil spills and ship traffic. Even though the PAHs are removed during wastewater treatment processes, mostly via sorption to sludge, their residues still remain in wastewater treatment plant (WWTP) effluents. Therefore, municipal and industrial WWTPs have been identified as the major point sources of water pollution by PAHs.

VOCs are a diverse family of carbon-based compounds that have high vapor pressures (>0.01 kPa at 293.15 K and 101.3 kPa), and therefore readily evaporate to the atmosphere. VOCs are released in natural processes (forest fires, vegetations) and anthropogenic activities. They may undergo complex chemical reactions in the atmosphere, causing a number of indirect effects, in particular the formation of ozone and fine particles. VOCs are contaminants of concern because of very large environmental releases, human toxicity, and a tendency for some compounds to persist in and migrate with groundwater to drinking water supply wells (Zogorski et al., 2006). Sources of VOCs in wastewater are industries (synthetic organic chemical manufacturing, plastics, pharmaceutical and pesticide production, etc.), commercial establishments (hospitals, commercial laundries), household and consumer products (cleaning products, disinfectants, personal hygiene compounds, etc.), and surface runoff, especially in very urbanized areas. VOCs can be formed during in-sewer biochemical processes and during wastewater treatment processes such as chlorination of wastewater effluents (chloroform and methylene chloride formation). In WWTPs, these volatile compounds can easily be released from wastewater to the atmosphere by volatilization and gas stripping in aerated treatment units, thereby causing potential adverse health effects, especially to the workers at the plants who are directly exposed to the toxic chemicals.

In this chapter, we will review some recent reports in the literature on the occurrence, removal, and fate of commonly studied polycyclic aromatic hydrocarbons and volatile organic compounds in municipal wastewater treatment plants.

5.1 PAHs in Municipal Wastewater Treatment Plants

Out of hundreds of PAHs identified in the environment, 16 PAHs are the most commonly studied, as they were named priority organic pollutants by the U.S. Environmental Protection Agency (USEPA) and the EU Water

Framework Directive (WFD). These PAHs are two-ring naphthalene; three-ring compounds acenaphthylene, acenaphthene, fluorene, phenanthrene, and anthracene; four-ring compounds fluoranthene, pyrene, benzo[a]anthracene, and chrysene; five-ring compounds benzo[a]pyrene, benzo[b]fluoranthene, benzo[k]fluoranthene, and dibenz[a,h]anthracene; and six-ring compounds indeno[1,2,3-c,d]pyrene and benzo[g,h,i]perylene. There have been numerous studies published about PAHs in water and wastewater treatments, and most of them focused on industrial wastewater treatments. In this chapter, we will highlight the results of the studies on the PAH presence in municipal WWTPs.

5.1.1 Occurrence of PAHs in Wastewater and Sludge

A wide range of the concentrations of PAHs has been reported for different WWTPs. The individual PAHs have been found at concentrations in the ng/L to low μg/L range in raw wastewater, while their concentrations ranged up to a few hundreds of ng/L in WWTP effluents. Figure 5.1 summarizes some recent data in the literature on the occurrence of the 16 PAHs in wastewater treatment system influent and effluent.

In general, low molecular weight PAHs (i.e., two- and three-ring PAHs), naphthalene, phenantrene, and fluoranthene, are reported to be present in highest concentrations in both the wastewater influent and effluent (Blanchard et al., 2004; Fatone et al., 2011; Tian et al., 2012). This is probably due to their higher water solubility compared with high molecular weight PAHs (i.e., four-, five-, and six-ring PAHs). Naphtalene was reported to be the most ubiquitous and abundant PAH in wastewater (Manoli and Samara, 2008; Fatone et al., 2011; Tian et al., 2012; Sun et al., 2013; Wang et al., 2013), which is also due to its wide commercial use in dusting powders, bathroom products, deodorant discs, wood preservatives, fungicides, and mothballs, and its use as an insecticide (Fatone et al., 2011). The International Agency for Research on Cancer determined fluoranthene and phenanthrene as "not classifiable" regarding their carcinogenicity to humans (IARC, 2002).

Although it has been classified as a noncarcinogen, fluoranthene might be an important contributor to the risk from PAH exposure because of its abundance in vehicle and other types of emissions, and also because the dietary intake of PAHs comprises relatively large amounts of this congener compared with other measured PAHs (Boström et al., 2002). Among high molecular weight PAHs, the four-ring pyrene and chrysene and five-ring benzo[a]pyrene and indeno[1,2,3-c,d]pyrene are the most commonly detected (Pham and Proulx, 1997; Blanchard et al., 2004; Tian et al., 2012; Wang et al., 2013). The IARC has determined benzo[a]pyrene and indeno[1,2,3-c,d]pyrene as "probably carcinogenic" and "possibly carcinogenic" to humans, respectively, while chrysene and pyrene are not classifiable regarding their carcinogenicity to humans (IARC, 2002). It has been shown that the carcinogenic and mutagenic activity of a PAH is associated with its structural features and the complexity

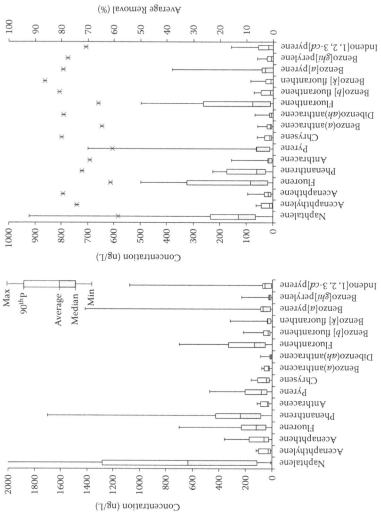

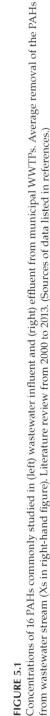

FIGURE 5.1

Concentrations of 16 PAHs commonly studied in (left) wastewater influent and (right) effluent from municipal WWTPs. Average removal of the PAHs from wastewater stream (Xs in right-hand figure). Literature review from 2000 to 2013. (Sources of data listed in references.)

of molecule, in such a way that a more complex compound is usually more potent (Boström et al., 2002).

PAHs typically occur in mixtures, and therefore, their concentrations are usually reported as the sum of analyzed congeners. A number of epidemiologic studies have shown that the mixtures of different PAHs are associated with an increased risk for tumor development. The sum of the concentrations of different PAH congeners will depend on the source and quality of raw wastewater, and the concentration ratio of individual PAHs in a mixture can be used to reveal their sources as each source generates a characteristic PAH pattern (Yunker et al., 2002). The total PAH concentrations have been reported to be up to a few µg/L in WWTP influents. The highest total PAH concentrations were typically reported for the municipal wastewater with a high contribution of industrial wastewater. Busetti et al. (2006) analyzed 16 USEPA PAHs at different points along the process line at the Fusina WWTP (Venice, Italy), which treats approximately 90,000 m³/day of domestic wastewater, a portion of the local urban runoff (atmospheric deposition and road traffic emissions deposition), approximately 10,000 m³/day of pretreated industrial wastewater from an industrial area, and approximately 300 m³/day of untreated industrial waste and sewage from ships. Of the summed concentrations of the 16 most commonly studied PAHs (Σ16PAHs), the highest levels—considering both the aqueous and solid fraction of samples—were found at the pretreated industrial influent from the sewage main serving a power station (4.62 µg/L) and in the interstitial water from the filtration of the secondary sludge (3.77 µg/L). Average concentrations of Σ16PAHs in the wastewater train were reduced to 1.12 µg/L, out of which only 5% was detected in the aqueous fraction of the effluent samples. Wang et al. (2013) analyzed 16 PAHs in wastewater treated by the WWTP of the city of Hefei (Anhui Province, East China), which treats 300,000 m³/day of wastewater whose origins are domestic (60%), industrial (automobile manufacturers, chemical and food industries), and landfill leachate. The wastewater is treated in three parallel activated sludge wastewater treatment systems (three anaerobic tanks followed by three oxidation ditches). The average Σ16PAH concentration of 5.7 µg/L was detected in the WWTP influent, and 2.2 µg/L in the WWTP effluent, with naphthalene, benzo[a]pyrene, and indeno[1,2,3-c,d]pyrene having the highest concentrations at both sampling points. Blanchard et al. (2004) determined the concentrations of PAHs and calculated the annual balances between different inflows (i.e., five sewers) and sludge of the Seine Aval WWTP in the Paris region (France). The WWTP serves around 8 million inhabitants and treats around 2,600,000 m³/day of domestic and industrial wastewater, with a fraction of runoff water from roadways. The Σ16PAH concentrations in both the dissolved and particulate phases of the WWTP influent were up to 28 µg/L, with 63% being sorbed to particulate matter. Two- and three-ring PAHs accounted for 80% of the total load, and five compounds, namely, phenanthrene, naphthalene, fluoranthene, pyrene, and chrysene, comprised 92%

of the total PAHs at the influent. They estimated that out of 26.7 kg/year of Σ16PAHs that reach the WWTP, about 0.53 kg/year was discharged by effluent wastewater to the river Seine. Manoli and Samara (2008) conducted a study on the WWTP of Thessaloniki in northern Greece, which treats domestic wastewater, a part of the local urban runoff, and a minor contribution (<5%) of industrial discharge in an activated sludge treatment process. Lower Σ16PAH concentration levels were reported by Tian et al. (2012) for a WWTP of Tai'an City (eastern China) that treats 50,000 m^3/day of municipal wastewater, with approximately 40% coming from industries (wineries, fiberglass factories, print works, and knitwear mills). The Σ16PAH concentration was around 1.1 μg/L in the WWTP influent, with naphthalene, fluorene, fluoranthene, and pyrene accounting for 53 to 90% of the total PAHs. Similar ranges of Σ16PAH concentrations were reported by Vogelsang et al. (2006) for five Norwegian WWTPs treating from 10,000 to 120,000 m^3/day of domestic wastewater. The total concentration of 16 PAHs in the five influents ranged from 0.2 to 1.3 μg/L, with the two- and three-ring PAHs typically constituting 60 to 80% of the total PAHs. Fatone et al. (2011) studied five typical Italian municipal WWTPs with treatment capacities of 12,000 to 700,000 population equivalent. The average total concentrations of 16 PAHs ranged from 0.14 to 1.54 μg/L for the WWTP influents, even for the WWTPs serving large and highly urbanized areas. The influent concentrations were reduced to 0.08 to 0.2 μg/L at the WWTP effluents. Napthalene was ubiquitous and most abundant in all the samples. Qiao et al. (2014) studied the occurrence and fate of PAHs and their transformation products in a WWTP, located in the northwest of Beijing (China), during summer (July) and winter (February–March) months. The WWTP serves a population of 814,000 and treats municipal wastewater in two parallel biological processes of 200,000 m^3/day capacity. The authors detected 372 ± 56.6 ng/L and 4260 ± 866 ng/g of total PAHs in the dissolved phase and suspended solids of influent samples, respectively, during summer, and 749 ± 69.9 ng/L and 5781 ± 622 ng/g during winter months. The authors calculated that the total mass load of PAHs discharged with the effluent, during winter, was 0.46 mg/day per capita, which is similar to the results for the WWTP in Thessaloniki: 0.36 mg/day per capita (Manoli and Samara, 1999), but higher than for the WWTP at Hefei in eastern China: 0.15 mg/day (Wang et al., 2013). Besides Qiao et al. (2014), several other studies have shown temporal variation in PAH concentrations. In general, higher PAH concentrations were observed for winter months, probably due to the emissions from residence heating systems or cold start of vehicles, and higher atmospheric deposition in the cold period (Pham and Proulx, 1997; Manoli and Samara, 1999; Blanchard et al., 2004; Wang et al., 2011).

Because of their lipophilic nature, PAHs preferentially partition onto particulate matter of wastewater and sludge during wastewater treatment. The fraction of PAHs sorbed to particulate matter has been reported to be in the range of 63 to 95% for WWTP influent and in the range of 50 to 95% for WWTP effluent (Blanchard et al., 2004; Busetti et al., 2006; Fatone et al., 2011;

Qiao et al., 2014). Qiao et al. (2014) estimated that 68% of the total PAH output load is discharged from the WWTP with sludge. The assessment of their concentrations in sludge is especially important when the sludge solids are applied to agricultural soil to improve its structure and fertility. PAHs may remain in the amended soil for months due to their sorption to organic, mineral, and amorphous phases of soil and slow biodegradation, and can thereby be taken up by plants and animals and accumulate in the terrestrial food chain (Stevens et al., 2003). Average Σ16PAH concentrations measured in different studies have been reported to range from around 0.03 mg/kg d.w. in Jordan (Jiries et al., 2000) to 50 mg/kg d.w. in the UK (Jones and Northcott, 2000; Stevens et al., 2003). The information about the concentration of PAHs in sludge, as reported in the referenced literature, is given in Table 5.1. In the EU, the third draft of the "Working Document on Sludge" of the European Commission proposed that the sum of nine PAH congeners—acenaphthene, phenanthrene, fluorene, fluoranthene, pyrene, benzo[b+j+k]fluoranthene, benzo[a]pyrene, benzo[g,h,i]perylene, and indeno[1,2,3-c,d]pyrene—should not exceed 6 mg/kg d.w. In the studies conducted not only in the UK (Jones and Northcott, 2000; Stevens et al., 2003) and Spain (Pérez et al., 2001; Villar et al., 2006), but also at locations as geographically disparate as China (Cai et al., 2007), Turkey (Ozcan et al., 2013), and Canada (Kohli et al., 2006), the EU Σ9PAH concentrations exceed the proposed limits even when they originate from domestic wastewater treatment plants. Jones and Northcott (2000) therefore concluded that implementation of this standard would result in the cessation of a very substantial proportion of sewage sludge applications to agricultural land in the EU. As for the individual PAHs, Table 5.1 summarizes some literature data on their occurrence in WWTP sludge. As shown in Table 5.1, the concentrations of individual PAHs generally range from the low µg/kg to a few mg/kg d.w. The most abundant and frequently detected PAHs in sludge are the three-ring phenanthrene and fluoranthene (Lazzari et al., 2000; Pérez et al., 2001; Miège et al., 2003; Kohli et al., 2006; Villar et al., 2006), followed by pyrene (Lazzari et al., 2000; Miège et al., 2003) and benzo[g,h,i] perylene (Miège et al., 2003; Villar et al., 2006).

5.1.2 Removal and Fate of PAHs during Municipal Wastewater Treatment

As previously observed, PAHs are hydrophobic compounds that show high affinity for organic matter. They are only partially removed during wastewater treatment, with an average removal of 60 to 90% from water, including their transfer to sludge (Figure 5.1). In the studies reviewed, removal reflects a transformation of the target PAHs to another, not measured, compound during wastewater treatment. Individual PAHs differ in their physicochemical properties that have a significant impact on polarity, solubility, and other properties that govern their fate during wastewater treatment. Actually, PAHs are characterized by relatively low water solubility, which depends upon temperature, pH, ionic strength, and water matrix components

TABLE 5.1

Concentration of PAHs Reported in Sludge from Municipal WWTPs

Compound	Concentration (µg/kg dry weight)[a]											
	Jordan	UK	Canada	Italy	Turkey	Turkey	Spain	Spain	China	China	China	China
Naphtalene	1.1–3.0	150–19,000	30–4700	SS: 31.5 TS: 28.4		89–231; 599–1506	27–287	PS: 230–1583[b] SS: 364–2515 TS: 322–1169	nd–2900	PS: 35.5 SS: 89.5 TS: 41.2	PS: 21.85 SS: 98.7 TS: 768.05	Inf SS: 301–571[c] EffSS: 1702 SS: 251
Acenaphthylene	0.1–0.4	30–100	10–11,000	SS: 52.5 TS: 67.2		17–344; 10–106	31–118	PS: 897 SS: 0 TS: 596–608	nd–1000	PS: 110.5 SS: 275.2 TS: 127.6	PS: 43.7 SS: 3.55 TS: 114.9	Inf SS: 32.6–34.7 EffSS: 507 SS: 83.2
Acenaphthene		1700–6600	10–1570	SS: 63 TS: 92.1		9–28; 25–317	149–492	PS: 121–438 SS: 121–470 TS: 163–509	nd–780	PS: 236.5 SS: 1532.4 TS: 432.8	PS: 5.6 SS: 4.45 TS: 103.1	Inf SS: 30.2–65. EffSS: 937 SS: 103.1
Fluorene	0.5–3.5	3600–8100	20–2580	SS: 75 TS: 73.8		16–186; 7–387	28–704	PS: 68–258 SS: 81–523 TS: 135–352	80–4800	PS: 76 SS: 1060.5 TS: 225.6	PS: 42 SS: 21.95 TS: 253.8	Inf SS: 168–357 EffSS: 6786 SS: 263
Phenanthrene	3.0–7.7	3100–16,000	40–10,400	SS: 81.3 TS: 91.1	nd	287–683; 58–2176	540–2030	PS: 313–1180 SS: 200–1271 TS: 429–1013	40–6600	PS: 285.6 SS: 1525 TS: 492	PS: 29.65 SS: 30.1 TS: 670.55	Inf SS: 716–1499 EffSS: 3584 SS: 1107
Anthracene	0.4–3.9	380–1800	30–17,500	SS: 90.3 TS: 78.8	0–127	45–867; 46–2037	34–234	PS: 26–66 SS: 24–72 TS: 34–92	nd–6100	PS: 15.5 SS: 271 TS: 65.5	PS: 5.8 SS: 2.9 TS: 100.25	Inf SS: 99–161 EffSS: 1866 SS: 132

(*Continued*)

TABLE 5.1 (Continued)

Concentration of PAHs Reported in Sludge from Municipal WWTPs

Compound	Jordan	UK	Canada	Italy	Turkey	Turkey	Spain	Spain	China	China	China	China
							Concentration (μg/kg dry weight)[a]					
Pyrene	4.2–6.2	2100–5600	30–15,000		255.9–519.6	323–821; 201–2077	277–702	PS: 578–2304 SS: 628–2502 TS: 824–1911	80–4300	PS: 39.8 SS: 482 TS: 109.8	PS: 25 SS: 5.5 TS: 247.7	Inf SS: 451–467 EffSS: 7168 SS: 302
Chrysene	1.2–1.9	1000–6000	70–20,300	SS: 129 TS: 94	299.8–543	nd–582; 2–1458	79–312	PS: 194–1101 SS: 238–913 TS: 101–1003	nd–4000	PS: 32 SS: 325 TS: 75	PS: 7.55 SS: 0.1 TS: 103.6	Inf SS: 217–269 EffSS: 1732 SS: 234
Benzo[a] anthracene	nd–2.0	600–2800	40–18,300	SS: 133.3 TS: 98.6	333.9–711.9	nd–483; 9–884	80–184	PS: 251–617 SS: 305–666 TS: 312–1413	13–11,000	PS: 34.3 SS: 32.4 TS: 78.6	PS: nd SS: nd TS: nd	Inf SS: 198–219 EffSS: 1603 SS: 176
Dibenzo[a,h] anthracene	nd–0.4	60–380	20–3410	SS: 97 TS: 52.3	769–1610	nd–60; nd–11	2–66	PS: 63–72 SS: 74–361 TS: 99–483	nd–220	PS: 8.7 SS: 27 TS: 10.8	PS: nd SS: nd TS: nd	Inf SS: 81.6–119 EffSS: 1562 SS: 64.6
Fluoranthene	1.4–4.8	1400–7400	30–17,500	SS: 94.6 TS: 93.1	92.9–236.6	115–679; 105–2445	100–629	PS: 119–278 SS: 94–298 TS: 193–518	nd–8000	PS: 174 SS: 470.6 TS: 215.4	PS: 41.4 SS: 13.1 TS: 375.45	Inf SS: 543–621 EffSS: 1122 SS: 394
Benzo[b] fluoranthene	nd–2.1	1100–7200	60–20,600	SS: 105 TS: 93.2	630.6–1109	nd–260; nd–836	35–479	PS: 104–637 SS: 122–449 TS: 129–457	nd–2900	PS: 8.8 SS: 39.3 TS: 13.8	PS: 5.45 SS: 0.25 TS: 88.65	Inf SS: 203–351 EffSS: 1765 SS: 285

(Continued)

TABLE 5.1 (*Continued*)

Concentration of PAHs Reported in Sludge from Municipal WWTPs

Compound	Concentration (µg/kg dry weight)[a]											
	Jordan	UK	Canada	Italy	Turkey	Turkey	Spain	Spain	China	China	China	China
Benzo[k]fluoranthene	0.3–1.0	700–4500	20–16,500	SS: 93 TS: 77.5	578–1090	nd–310; nd–354	11–289	PS: 36–93 SS: 45–65 TS: 41–90	7–7000	PS: 1.02 SS: 18.2 TS: 4.1	PS: 0.45 SS: 0.3 TS: 26.1	Inf SS: 85–175 EffSS: 1461 SS: 109
Benzo[a]pyrene	0.5–1.9	690–400	20–19,600	SS: 97 TS: 86.8	706.9–1172	23–524; 14–174	23–522	PS: 65–437 SS: 39–210 TS: 52–340	7–6600	PS: 2.86 SS: 34.9 TS: 8.76	PS: nd SS: nd TS: nd	Inf SS: 202–766 EffSS: 1570 SS: 556
Benzo[g,h,i]perylene	4.7–7.3	0470–2300	20–13,500	SS: 84 TS: 68.9	766.6–1584	nd–38; nd–1829	21–589	PS: 298–2255 SS: 203–1127 TS: 358–1219		PS: 0.16 SS: 32.05 TS: 5.7	PS: nd SS: nd TS: nd	Inf SS: 215–316 EffSS: 1784 SS: 281
Indeno[1,2,3-c,d]pyrene	nd–1.4	390–2700	30–12,000		840.7–1797	nd–119; nd–59	189–461	PS: 64–257 SS: 73–248 TS: 46–202	nd–480	PS: 5.59 SS: 16.3 TS: 7.6	PS: nd SS: nd TS: nd	Inf SS: 242–267 EffSS: 2084 SS: 271
No. of WWTPs	3	14	35	2	2	2	2		9			
Source	Jiries et al., 2000	Stevens et al., 2003	Kohli et al., 2006	Busetti et al., 2006	Salihoglu et al., 2010	Ozcan et al., 2013	Pérez et al., 2001	Villar et al., 2006	Cai et al., 2007	Yao et al., 2012	Tian et al., 2012	Qiao et al., 2014

Note: nd, not detected; PS, primary sludge; SS, secondary sludge; TS, treated (stabilized) sludge; InfSS and EffSS, influent and effluent suspended solids, respectively.

a Range: min–max for all the WWTPs.
b Range for 12 campaigns.
c Range summer–winter.

(i.e., dissolved organic carbon) (Dabestani and Ivanov, 1999). The solubility of PAHs in water ranges from 31 mg/L for two-ringed naphthalene to 0.00019 mg/L for six-ringed indeno[1,2,3-c,d]pyrene; low molecular weight (LMW) PAHs (<200 g/mol) are much more water soluble than high molecular weight (HMW) congeners, which tend to sorb to particulate matter during wastewater treatment (Lundstedt, 2003). Therefore, the LMW PAHs are partly dissolved, and thus highly available for various microbial degradation/transformation processes, while the HMW PAHs are primarily associated with solids in water and less available for degradation (Lundstedt, 2003; Bergqvist et al., 2006). The principal removal mechanism for HMW PAHs is therefore via sorption to sludge particles and transfer to the sludge processing systems and, to a certain extent, removal into the final effluent associated with suspended solids (Byrns, 2001). Since urban wastewater usually takes hours to reach WWTPs, the interactions of PAHs with suspended particulates are likely to be in equilibrium by the time the wastewater enters the WWTP headwork (Fatone et al., 2011). Sorption was found to be the main removal mechanism of more hydrophobic HMW PAHS in the primary sedimentation tank (Manoli and Samara, 2008). Manoli and Samara (1999, 2008) reported the aqueous removal of PAHs during primary treatment as ranging from 28% for naphtalene to 67% for benzo[a]anthracene. Five- and six-ring PAHs exhibited much higher aqueous removal in primary treatment (51 to 56%) than in secondary treatment (ranging from <1 to 20%), and the overall removal of PAHs was calculated to range from 37% for fluorene to 89% for benzo[a]anthracene, with the five- and six-ring PAHs being removed by 55 to 75%. Wang et al. (2013) reported an overall average removal of 61% (from 29% for phenanthrene to 86% for dibenzo[a,h]anthracene), where the primary sedimentation contributed 35% of the overall aqueous removal. Also, Tian et al. (2012) observed that the highest reduction in PAH concentrations occurred during their pass through the grit chamber and primary sedimentation; i.e., the average removal of 12 PAHs was 76% of their total concentration at the influent.

In biological wastewater treatment processes, and especially under aerobic conditions, PAHs are subject to a combined effect of microbial biodegradation/transformation, biosorption, and volatilization. Sorption of PAHs to biomass is complex, and it possibly involves both active surface reactions and nonspecific partitioning phenomena (Stringfellow and Alvarez-Cohen, 1999). LMW PAHs are to a certain extent more easily biodegraded than the HMW PAHs, as the sorption of PAHs to biomass may inhibit their biodegradation. However, PAH biosorption may have a positive impact on the removal of PAHs during activated sludge treatment, as sorbed PAHs are retained in the treatment system for a longer period than the hydraulic residence time, so that the PAH-degrading bacteria have a longer period in which to act upon the PAH (Stringfellow and Alvarez-Cohen, 1999). The concentrations of total PAHs coming from primary treatments are reduced by up to 60% during activated sludge processes, with, in general, lower (<40%) removal of five- and six-ring PAHs (Manoli and Samara, 2008; Tian et al., 2012; Yao et al., 2012; Qiao et al., 2014). The aeration in the activated sludge process may lead to the transfer of PAHs to the atmosphere, which mainly depends on the volatility (i.e., Henry's law constant) of the compounds. The vapor pressure of PAHs at 25°C ranges from 10.4 Pa for two-ringed naphthalene to 1.35×10^{-13} Pa for six-ringed indeno[1,2,3-c,d]pyrene, indicating a tendency for the LMW PAHs to be found in higher concentrations in the air. Therefore, the air stripping in secondary aerobic treatment may be an important removal mechanism only for more volatile and more water-soluble low molecular weight PAHs (Manoli and Samara, 2008). Nevertheless, Kappen (2003) found that only a small percent of naphthalene (0.49%) and phenanthrene (0.01%) that reach a WWTP leave in the air stream.

Besides physicochemical properties of PAHs, WWTP operation factors, such as the composition of wastewater, technical setup (e.g., conventional activated sludge treatment or membrane bioreactor (MBR)), chemical/mechanical/biological compartments, and sludge solids concentration, largely influence their removal and fate during wastewater treatment. Therefore, a wide variation in removal efficiencies is reported for each individual PAH compound and across different treatment processes. Vogelsang et al. (2006) observed that the WWTPs with combined biological (anoxic and aerobic-activated sludge process) and chemical treatment provided generally better removal of Σ16PAH, i.e., 94 to 100%, in comparison with the chemical treatment plants that removed 61 to 78% of Σ16PAH. Biological/chemical treatment resulted in 91 to 100% removal of two- and three-ring PAHs, which may be attributed to biodegradation or evaporation of these PAHs. The plants that applied mechanical pretreatment and chemical treatment removed 82 to 100% of the highly lipophilic four-, five-, and six-ring PAHs (log P 4.9 to 6.8), but around 29 to 70% of the less lipophilic two- and three-ring PAHs. The mechanical treatment plant removed 25 to 26% of the Σ16PAH, without a significant difference in the removal of individual PAHs. Fatone et al. (2011) studied the removal of PAHs at five Italian WWTPs with different treatment capacities (i.e., 12,000 to 700,000 population equivalent), quality of raw wastewater, and types of biological processes (predenitrification-nitrification, extended oxidation, multizone biological nutrient removal, intermittent aeration of continuously fed bioreactors, MBRs) and reported the removal of 40 to 60% and 10 to 90% of the PAHs (after the removal at the WWTP headworks) during primary and secondary treatment in the WWTPs, respectively. The authors observed that the conventional activated sludge process (CASP) and pilot MBR systems, operating in parallel, provided similar overall removal of PAHs, i.e., 80 to 95%. The two systems operated with similar hydraulic retention times (HRTs), but the MBRs operated with much longer solid retention times, 200 to 500 days for MBR and 12 days for CASP. The high solid retention time of the MBRs improved the elimination of PAHs by sorption, but the biological activity decreased, as shown by (1) a lower volatile-to-total solids ratio with higher retention time in the MBR (i.e., decreased from 80 to 52%) and (2) a lower maximum specific oxygen utilization rate (sOUR) that decreased from 17 to 8 mg O_2/(g VSS-h) (Fatone et al., 2011).

5.2 VOCs in Municipal Wastewater Treatment Plants

There are thousands of compounds that belong to the class of VOCs, but most surveys focus on a hundred compounds containing 2 to 12 carbon atoms. Aliphatic hydrocarbons, aromatic hydrocarbons, halogenated volatiles, and dimethyl disulfide account for approximately 70% of all VOCs detected in municipal WWTPs (Rodriguez et al., 2012). The U.S. Clean Air Act (42 U.S. Code § 7412) addresses a large number of VOCs among 187 compounds recommended to be controlled by the USEPA as hazardous air pollutants, i.e., those pollutants that cause or may cause cancer or other serious health effects, such as reproductive effects or birth defects, or adverse environmental and ecological effects (EPA). In Europe, the Water Framework Directive (2000/60/EC) and the Directive on Environmental Quality Standards (Directive 2008/105/EC) of the European Commission address several VOCs in the list of priority compounds, which are benzene, C10–13-chloroalkanes, 1,2-dichloroethane, dichloromethane, hexachlorobutadiene, trichlorobenzenes, and trichloromethane. Table 5.2 reports the ranges of the concentrations of some commonly studied VOCs detected in samples of wastewater from sewer systems, and the influent and effluent of municipal wastewater treatment plants.

5.2.1 Occurrence of VOCs in the WWTP Influent and Effluent

VOCs are detected in raw municipal wastewater in ng/L to a few hundreds µg/L concentration range, and with markedly reduced concentrations at WWTP effluents. Cervera et al. (2011) analyzed 23 VOCs in the influent and effluent wastewater from an urban WWTP in Castellon (Spain). The highest concentrations at the WWTP influent were measured for toluene (179.7 µg/L), tetrachloroethylene (132.3 µg/L), and chloroform (96.0 µg/L), and at the WWTP effluent, for toluene (179.9 µg/L), o-xylene (87.4 µg/L), and chloroform (67.7 µg/L). Escalas et al. (2003) analyzed 47 volatile organic compounds, including 10 EU priority compounds (1,2,3-trichlorobenzene, 1,2,4-trichlorobenzene, carbon tetrachloride, chloroform, hexachlorobutadiene, trichloroethylene, perchloroethylene, 1,3,5-trichlorobenzene, naphtalene, benzene), at four sampling points of a municipal WWTP in Manresa (Spain). Only 6 out of 47 compounds were not detected. The authors observed high variations in the concentrations of the target analytes. Average mass flow of the detected compounds at the influent was 6380 g/day. Petroleum solvents and terpenic compounds from

TABLE 5.2

Overview of Concentrations' VOCs Reported for the In-Sewer Wastewater and the Municipal WWTP Influent and Effluent Wastewater

	Concentrations (μg/L)		
Compounds	In-Sewer Wastewater	Influent Wastewater	Effluent Wastewater
1,1,1-Trichloroethane		0.042–6.3	nd–0.26
1,2,3-Trichlorobenzene		nd–0.38	nd–0.28
1,2,4-Trichlorobenzene		nd–4.0	nd
1,2,4-Trimethylbenzene		0.168–113	0.066–1.2
1,2-Dichlorobenzene		<0.005	nd–3.0
1,2-Dichloroethane			0.12
1,2-Dichloropropane			0.2
1,3,5-Trimethylbenzene		nd–10	0.097
1,3-Dichlorobenzene		<0.005–0.04	<0.01–1.4
1,4-Dichlorobenzene		<0.005–1.054	<0.01–3.2
2-Chlorotoluene		<0.005–0.208	
4-Chlorotoluene		0.05–0.306	
Acrylonitrile			0.06
Benzene	<1.0	<0.005–32	nd–34.3
Bromobenzene		<0.005	
Bromodichloromethane			0.12–44.25
Bromoform		0.4	0.19–70.9
Carbon tetrachloride		0.18–4.4	0.13–0.26
Chlorobenzene		0.02–0.68	<0.01
Chloroform	<1.0–25	nd–96	2.9–67.7
cis-1,2-Dichloroethene		48	0.4
Dibromochloromethane		0.3	0.14–80.31
Dichloroethane	<1.0		
Dichloromethane	<1.0–86		0.2
Ethylbenzene		0.093–8.5	nd–2.3
Hexachlorobutadiene	<1.0	nd	
Isopropylbenzene		0.063–0.53	
m-Xylene + p-xylene		0.15–55.3	0.1–53.6
n-Butylbenzene		<0.005–0.156	
n-Propylbenzene		<0.005–7.5	0.088
o-Xylene		0.035–67.6	nd–87.4
p-Isopropyltoluene		<0.005–0.56	
Sec-butylbenzene		nd–0.76	
Styrene		0.139–1.2	0.059
Tert-butylbenzene		<0.005	
Tetrachloroethylene	<1.0–4	nd–132.3	0.11–48

(Continued)

TABLE 5.2 *(Continued)*

Overview of Concentrations' VOCs Reported for the In-Sewer Wastewater and the Municipal WWTP Influent and Effluent Wastewater

	Concentrations (µg/L)		
Compounds	In-Sewer Wastewater	Influent Wastewater	Effluent Wastewater
Toluene	<1.0–3.2	3.008–193.3	0.01–180
Trichloroethylene	<1.0–1.8	49.3	nd–39.9
References	Gasperi et al., 2008	Cervera et al., 2011; Escalas et al., 2003; Fatone et al., 2011	Barco-Bonilla et al., 2013a, 2013b; Cervera et al., 2011; Escalas et al., 2003; Martí et al., 2011; Rodriguez et al., 2012

surfactant-containing preparations accounted for 79%, while the compounds classified as carcinogenic, toxic, or mutagenic (CTM) to reproduction in Directive 1999/13/EC, i.e., chloroform, benzene, carbon tetrachloride, trichloroethylene, and perchloroethylene, accounted for 14.5% of the total mass flow of VOCs at the WWTP influent. Sixteen VOCs listed as hazardous air pollutants (HAPs) by the Clean Air Act (EPA) accounted for 35% or 2240 g/day. The authors concluded that the total HAP emission from the WWTP (i.e., influent yearly mass flow of 817 kg/year) was much lower than the limit set by the USEPA for WWTPs (i.e., 22,680 kg/year) (Code of Federal Regulations, USEPA). The total VOC concentrations were reduced by 89%, on average. A daily discharge via effluent wastewater was estimated to be 681 g/day, distributed as follows: decane (<0.5 g/day), chlorinated aliphatic hydrocarbons (125 g/day), light aromatic hydrocarbons (391 g/day), halogenated aromatic hydrocarbons (6.1 g/day), sulfur compounds (78 g/day), and terpenic compounds (81 g/day). Fatone et al. (2011) analyzed 23 VOCs and found BTEX compounds (excluding benzene), 1,2,4-trimethylbenzene, and 4-chlorotoluene to be the most commonly detected VOCs in five Italian municipal WWTPs. For benzene, toluene, ethylbenzene, and xylene (BTEX), compounds were presumed to result from vehicle emissions (e.g., exhaust, fuel evaporation). Toluene was detected in the highest concentrations, ranging from 3 to 7 µg/L, while all the other compounds had concentrations lower than 1 µg/L. At the WWTP effluents, toluene was again the most abundant VOC (0.5 to 2.7 µg/L). The compounds studied showed low sorption affinity, as their fraction in the suspended particulate matter of the WWTP influents was always below method detection limits (0.5 mg/kg TS). In a study by Barco-Bonilla et al. (2013b), the concentrations of VOCs at the effluent of 12 WWTPs in Almeria Province (Spain) ranged from 0.10 µg/L (trichloroethylene) to 0.81 µg/L (m-xylene + p-xylene). Toluene, o-xylene, and m-xylene + p-xylene were detected in all the samples. The authors noted that the concentrations of VOCs in the samples collected in July were lower than in those taken during the spring months, which was attributed to their increased volatilization during summer. Similar seasonal differences were observed in a study of Rodriguez et al. (2012). They conducted the research at three municipal WWTPs and two microfiltration (MF)/reverse osmosis (RO) plants in Perth (Australia) to determine the concentration of VOCs in secondary treated effluent and post-reverse osmosis treatment, and to assess the health risk associated with VOCs for indirect potable reuse. Out of 61 VOCs, 21 (34% of the total) were in the effluent of the three municipal WWTPs, with 1,4-dichlorobenzene (93.9%), tetrachloroethene (87.9%), carbon disulfide (81.2%), and chloromethane (57.6%) as the most frequently detected VOCs. The authors observed that the concentrations varied widely for different compounds in different WWTPs. The concentrations of cis-1,2-dichloroethene, chloromethane, tetrachloroethene, and trichloroethene in the Kwinana Water Reclamation Plant (KWRP) influent were statistically higher than in the WWTP Beenyup secondary effluent, which may be related to the fact that WWTP KWRP is on the site of an oil refinery. The authors highlighted the importance of wastewater characterization if indirect potable reuse is intended, given the different nature of industry, trade waste agreements/regulations, sewer arrangements, and WWTP processes in place (Rodriguez et al., 2012). These results are consistent with other studies that pointed out that the VOC concentrations in wastewater are closely related to industrial activities in the catchment (Cheng et al., 2008; Fatone et al., 2011; Martí et al., 2011). The concentrations of VOCs in sewer systems are scarcely reported in the literature. Gasperi et al. (2008) chose the Paris combined sewer as a representative of typical European sewerage systems, and studied 66 compounds, including 7 metals (Cd, Cr, Cu, Hg, Ni, Pb, and Zn) and 59 organic compounds (organotins, chlorobenzenes, VOCs, PAHs, pesticides, etc.) during dry and wet weather periods, and on combined sewer overflows. During dry weather periods, chloroform and tetrachloroethylene were detected with highest frequency in raw wastewater at median total concentrations of 8.5 and 1.6 µg/L, respectively, including both the dissolved and particulate fractions. Similar levels of VOCs were reported for 30 WWTPs in the UK (Rule et al., 2006). During wet weather periods, chloroform, trichloroethylene, and tetrachloroethylene were detected in most samples, with concentrations ranging from 1 to 58 µg/L. Authors indicated that considering that VOCs easily evaporate, the gas phase washout by rain, which is generally considered a major mechanism for removal from the atmosphere, could appear as a major contributor to VOC loads in combined sewer overflow.

5.2.2 Removal and Fate of VOCs during Wastewater Treatment

As in the previous sections, the term *removal* is used to describe the VOC reduction from the influent to the effluent of a WWTP, due to a combination of different processes: chemical and physical transformation, biodegradation, volatilization, or sorption to solid matter. In general, very high VOC removal was reported for typical municipal wastewater treatment conditions (Escalas et al., 2003; Fatone et al., 2011). Out of 23 VOCs studied by Fatone et al. (2011), only toluene was still present in five WWTP effluents, probably due to its high concentrations in the raw wastewaters. The total VOC removal varied during different treatment sections, i.e., 40 to 60% during pretreatment and primary treatments, and 10 to 50% of the remaining VOCs (i.e., after their removal in the headwork) during secondary biological treatments. The MBR technology did not provide any significant advantage over conventional treatments in terms of VOC removal. Rodriguez et al. (2012) observed very high variability in the removal of different VOCs from secondary effluent to post-RO treatment. The median removal efficiency ranged from −77.0% for dichlorodifluoromethane to 91.2% for tetrachloroethene, and for 10 of the 21 compounds detected in secondary effluent, the median removal was above 70%. Escalas et al. (2003) observed an increase in the total VOC concentrations from the WWTP influent (187 μg/L) to the primary treatment (215 μg/L), which they attributed to the recirculation of the supernatant from anaerobic digestion to the headwork. After primary treatment, the total VOC concentrations decreased, and the average overall removal in the WWTP was around 90%. Kemp et al. (2000) evaluated the impact of operating conditions, i.e., anaerobic, anoxic denitrifying, and aerobic, on biotransformation rates of VOCs and found minimal biotransformation of the target VOCs under anaerobic and anoxic denitrifying conditions. In the aerobic environment, they observed higher biotransformation rates for the three nonchlorinated VOCs: toluene, o-xylene, and 1,3,5-trimethylbenzene, compared to the chlorinated VOCs.

VOCs may be removed by various chemical, physical, and biological processes during wastewater treatment. In WWTPs, photodegradation is not expected to be a significant removal mechanism due to the opacity of the system, the depth of the liquid, and the residence time of the processes. Volatile organics do not have a high affinity for wastewater solids, and thus adsorption is not expected to be a significant removal mechanism, except for large loadings of suspended solids and oils in the wastewater. The major VOC removal mechanisms include volatilization and biodegradation, and one or the other mechanism will prevail, depending on the particular compound properties and design and operational parameters of the biological treatment process (EPA, 1994). Interrelated design and operating parameters that impact the major removal pathways include the dissolved oxygen (DO), mixed liquor suspended solids concentration, organic loading rate (OLR), and solids retention time (SRT), where maximum biodegradation and minimum volatilization removal rate are observed for long SRT, high OLR, and low DO (Min and Ergas, 2006). Biodegradation can reduce the concentrations of VOCs in WWTPs to a certain degree, especially of the easily biodegradable VOCs such as esters, benzene, toluene, and phenols, but the atmospheric emissions in the aeration basins as well as in different treatment units generally lead to significantly lower dissolved concentrations in the WWTP effluent (Fatone et al., 2011). The aeration during wastewater treatment and during sludge treatment processes can achieve more than 90% removal of the VOC concentration in raw wastewater (NRC, 1996). For example, Wu et al. (2002) reported a 96% decrease in total VOCs in a WWTP during exposure to the atmosphere via air stripping. Chern and Chou (1999) observed that (1) for constant water temperature and air humidity, the VOC emission rates from batch aeration tanks and continuous-flow tanks increase linearly with increasing air temperature, especially for VOCs with lower Henry's law constants; and (2) for constant water and air temperatures, the VOC emission rates from batch aeration tanks and continuous-flow tanks increase with increasing relative humidity, again, especially for VOCs with lower Henry's law constants. Air emission prevails in pretreatment and primary clarifiers (Escalas et al., 2003). Yang et al. (2012) examined the VOC emissions and the related cancer and noncancer risks from inhalation intake, and observed that primary sedimentation is the treatment unit with the highest VOC emission and health risk to the workers. During aerobic treatment processes, which are characterized by high VOC emission, the associated health risks have been found to be limited due to the low VOC concentrations in the gas phase, as a result of the strong mixing and dilution with fresh air.

References

Barco-Bonilla, N., Romero-Gonzalez, R., Plaza-Bolanos, P., Martinez Vidal, J.L., Castro, A.J., Martin, I., Salas, J.J., Frenich, A.G. 2013a. Priority organic compounds in wastewater effluents from the Mediterranean and Atlantic basins of Andalusia (Spain). *Environmental Science: Processes and Impacts* 15, 2194–2203.

Barco-Bonilla, N., Romero-González, R., Plaza-Bolaños, P., Martínez Vidal, J.L., Garrido Frenich, A. 2013b. Systematic study of the contamination of wastewater treatment plant effluents by organic priority compounds in Almeria Province (SE Spain). *Science of the Total Environment* 447, 381–389.

Bergqvist, P.A., Augulytė, L., Jurjonienė, V. 2006. PAH and PCB removal efficiencies in Umeå (Sweden) and Šiauliai (Lithuania) municipal wastewater treatment plants. *Water, Air, and Soil Pollution* 175, 291–303.

Blanchard, M., Teil, M.J., Ollivon, D., Legenti, L., Chevreuil, M. 2004. Polycyclic aromatic hydrocarbons and polychlorobiphenyls in wastewaters and sewage sludges from the Paris area (France). *Environmental Research* 95, 184–197.

Boström, C.E., Gerde, P., Hanberg, A., Jernström, B., Johansson, C., Kyrklund, T., Rannug, A., Törnqvist, M., Victorin, K., Westerholm, R. 2002. Cancer risk assessment, indicators, and guidelines for polycyclic aromatic hydrocarbons in the ambient air. *Environmental Health Perspectives* 110, 451–488.

Busetti, F., Heitz, A., Cuomo, M., Badoer, S., Traverso, P. 2006. Determination of sixteen polycyclic aromatic hydrocarbons in aqueous and solid samples from an Italian wastewater treatment plant. *Journal of Chromatography A* 1102, 104–115.

Byrns, G. 2001. The fate of xenobiotic organic compounds in wastewater treatment plants. *Water Research* 35, 2523–2533.

Cai, Q.-Y., Mo, C.-H., Wu, Q.-T., Zeng, Q.-Y., Katsoyiannis, A. 2007. Occurrence of organic contaminants in sewage sludges from eleven wastewater treatment plants, China. *Chemosphere* 68, 1751–1762.

Cervera, M.I., Beltran, J., Lopez, F.J., Hernandez, F. 2011. Determination of volatile organic compounds in water by headspace solid-phase microextraction gas chromatography coupled to tandem mass spectrometry with triple quadrupole analyzer. *Analytica Chimica Acta* 704, 87–97.

Cheng, W.-H., Hsu, S.-K., Chou, M.-S. 2008. Volatile organic compound emissions from wastewater treatment plants in Taiwan: Legal regulations and costs of control. *Journal of Environmental Management* 88, 1485–1494.

Chern, J.-M., Chou, S.-R. 1999. Volatile organic compound emission rates from mechanical surface aerators: Mass-transfer modeling. *Industrial and Engineering Chemistry Research* 38, 3176–3185.

Code of Federal Regulations. Toxics criteria for those states not complying with Clean Water Act Section 303(c)(2)(B). 40 CFR 131.36. U.S. Environmental Protection Agency.

Council Directive 1999/13/EC of March 11, 1999, on the limitation of emissions of volatile organic compounds due to the use of organic solvents in certain activities and installations.

Dabestani, R., Ivanov, I.N. 1999. A compilation of physical, spectroscopic and photophysical properties of polycyclic aromatic hydrocarbons. *Photochemistry and Photobiology* 70, 10–34.

Directive 2000/60/EC of the European Parliament and of the Council establishing a framework for the community action in the field of water policy—EU Water Framework Directive.

Directive 2008/105/EC of the European Parliament and of the Council of December 16, 2008, on environmental quality standards in the field of water policy.

EPA. Clean Air Act—Air quality and emission limitations. CAA 112 Code § 7412: Hazardous air pollutants. http://www.epa.gov/air/caa/.

EPA. 1994. Air emissions models for waste and wastewater, North Carolina, U.S. EPA-453/R-94-080A.

Escalas, A., Guadayol, J.M., Cortina, M., Rivera, J., Caixach, J. 2003. Time and space patterns of volatile organic compounds in a sewage treatment plant. *Water Research* 37, 3913–3920.

Fatone, F., Di Fabio, S., Bolzonella, D., Cecchi, F. 2011. Fate of aromatic hydrocarbons in Italian municipal wastewater systems: An overview of wastewater treatment using conventional activated-sludge processes (CASP) and membrane bioreactors (MBRs). *Water Research* 45, 93–104.

Gasperi, J., Garnaud, S., Rocher, V., Moilleron, R. 2008. Priority pollutants in wastewater and combined sewer overflow. *Science of the Total Environment* 407, 263–272.

IARC. 2002. Naphthalene. IARC Summary and Evaluation, vol. 82. International Agency for Research on Cancer, Lyon, France.

Jiries, A., Hussain, H., Lintelmann, J. 2000. Determination of polycyclic aromatic hydrocarbons in wastewater, sediments, sludge and plants in Karak Province, Jordan. *Water, Air, and Soil Pollution* 121, 217–228.

Jones, K.C., Northcott, G.L. 2000. Organic contaminants in sewage sludges: A survey of UK samples and a consideration of their significance. Final report to the Department of the Environment, Transport and the Regions Water Quality Division.

Kappen, L.L. 2003. Volatilization and fate of polycyclic aromatic hydrocarbons during wastewater treatment. Thesis, College of Engineering, University of Cincinnati, Cincinnati, OH.

Kemp, J., Zytner, R.G., Bell, J., Parker, W., Thompson, D., Rittmann, B.E. 2000. A method for determining VOC biotransformation rates. *Water Research* 34, 3531–3542.

Kohli, J., Lee, H.-B., Peart, T.E. 2006. Organic contaminants in Canadian municipal sewage sludge. Part II. Persistent chlorinated compounds and polycyclic aromatic hydrocarbons. *Water Quality Research Journal of Canada* 41, 47–55.

Lazzari, L., Sperni, L., Bertin, P., Pavoni, B. 2000. Correlation between inorganic (heavy metals) and organic (PCBs and PAHs) micropollutant concentrations during sewage sludge composting processes. *Chemosphere* 41, 427–435.

Lundstedt, S. 2003. Analysis of PAHs and their transformation products in contaminated soil and remedial processes. PhD thesis, Umeå University, Sweden.

Manoli, E., Samara, C. 1999. Occurrence and mass balance of polycyclic aromatic hydrocarbons in the Thessaloniki sewage treatment plant. *Journal of Environmental Quality* 28, 176–187.

Manoli, E., Samara, C. 2008. The removal of polycyclic aromatic hydrocarbons in the wastewater treatment process: Experimental calculations and model predictions. *Environmental Pollution* 151, 477–485.

Martí, N., Aguado, D., Segovia-Martínez, L., Bouzas, A., Seco, A. 2011. Occurrence of priority pollutants in WWTP effluents and Mediterranean coastal waters of Spain. *Marine Pollution Bulletin* 62, 615–625.

Miège, C., Dugay, J., Hennion, M.C. 2003. Optimization, validation and comparison of various extraction techniques for the trace determination of polycyclic aromatic hydrocarbons in sewage sludges by liquid chromatography coupled to diode-array and fluorescence detection. *Journal of Chromatography A* 995, 87–97.

Min, K., Ergas, S. 2006. Volatilization and biodegradation of VOCs in membrane bioreactors (MBR). *Water, Air, and Soil Pollution: Focus* 6, 83–96.

NRC. 1996. *Use of reclaimed water and sludge in food crop production.* National Academic Press, Washington, DC.

Ozcan, S., Tor, A., Aydin, M.E. 2013. Investigation on the levels of heavy metals, polycyclic aromatic hydrocarbons, and polychlorinated biphenyls in sewage sludge samples and ecotoxicological testing. *CLEAN—Soil, Air, Water* 41, 411–418.

Pérez, S., Guillamón, M., Barceló, D. 2001. Quantitative analysis of polycyclic aromatic hydrocarbons in sewage sludge from wastewater treatment plants. *Journal of Chromatography A* 938, 57–65.

Pham, T.-T., Proulx, S. 1997. PCBs and PAHs in the Montreal urban community (Quebec, Canada) wastewater treatment plant and in the effluent plume in the St Lawrence River. *Water Research* 31, 1887–1896.

Qiao, M., Qi, W., Liu, H., Qu, J. 2014. Occurrence, behavior and removal of typical substituted and parent polycyclic aromatic hydrocarbons in a biological wastewater treatment plant. *Water Research* 52, 11–19.

Rodriguez, C., Linge, K., Blair, P., Busetti, F., Devine, B., Van Buynder, P., Weinstein, P., Cook, A. 2012. Recycled water: Potential health risks from volatile organic compounds and use of 1,4-dichlorobenzene as treatment performance indicator. *Water Research* 46, 93–106.

Rule, K.L., Comber, S.D.W., Ross, D., Thornton, A., Makropoulos, C.K., Rautiu, R. 2006. Survey of priority substances entering thirty English wastewater treatment works. *Water and Environment Journal* 20, 177–184.

Salihoglu, N.K., Salihoglu, G., Tasdemir, Y., Cindoruk, S.S., Yolsal, D., Ogulmus, R., Karaca, G. 2010. Comparison of polycyclic aromatic hydrocarbons levels in sludges from municipal and industrial wastewater treatment plants. *Archives of Environmental Contamination and Toxicology* 58, 523–534.

Stevens, J.L., Northcott, G.L., Stern, G.A., Tomy, G.T., Jones, K.C. 2003. PAHs, PCBs, PCNs, organochlorine pesticides, synthetic musks, and polychlorinated n-alkanes in U.K. sewage sludge: Survey results and implications. *Environmental Science and Technology* 37, 462–467.

Stringfellow, W.T., Alvarez-Cohen, L. 1999. Evaluating the relationship between the sorption of PAHs to bacterial biomass and biodegradation. *Water Research* 33, 2535–2544.

Sun, H., Tian, W., Wang, Y. 2013. Occurrence and fate of polycyclic aromatic hydrocarbons in the anaerobic-anoxic-oxic wastewater treatment process. *Advanced Materials Research* 610–613, 1722–1725.

Tian, W., Bai, J., Liu, K., Sun, H., Zhao, Y. 2012. Occurrence and removal of polycyclic aromatic hydrocarbons in the wastewater treatment process. *Ecotoxicology and Environmental Safety* 82, 1–7.

Villar, P., Callejón, M., Alonso, E., Jiménez, J.C., Guiraúm, A. 2006. Temporal evolution of polycyclic aromatic hydrocarbons (PAHs) in sludge from wastewater treatment plants: Comparison between PAHs and heavy metals. *Chemosphere* 64, 535–541.

Vogelsang, C., Grung, M., Jantsch, T.G., Tollefsen, K.E., Liltved, H. 2006. Occurrence and removal of selected organic micropollutants at mechanical, chemical and advanced wastewater treatment plants in Norway. *Water Research* 40, 3559–3570.

Wang, W., Massey Simonich, S.L., Giri, B., Xue, M., Zhao, J., Chen, S., Shen, H., Shen, G., Wang, R., Cao, J., Tao, S. 2011. Spatial distribution and seasonal variation of atmospheric bulk deposition of polycyclic aromatic hydrocarbons in Beijing–Tianjin region, North China. *Environmental Pollution* 159, 287–293.

Wang, X., Xi, B., Huo, S., Sun, W., Pan, H., Zhang, J., Ren, Y., Liu, H. 2013. Characterization, treatment and releases of PBDEs and PAHs in a typical municipal sewage treatment plant situated beside an urban river, East China. *Journal of Environmental Sciences* 25, 1281–1290.

Wu, C.H., Lu, J.T., Lo, J.G. 2002. Analysis of volatile organic compounds in wastewater during various stages of treatment for high-tech industries. *Chromatographia* 56, 91–98.

Yang, W.-B., Chen, W.-H., Yuan, C.-S., Yang, J.-C., Zhao, Q.-L. 2012. Comparative assessments of VOC emission rates and associated health risks from wastewater treatment processes. *Journal of Environmental Monitoring* 14, 2464–2474.

Yao, M., Zhang, X., Lei, L. 2012. Polycyclic aromatic hydrocarbons in the centralized wastewater treatment plant of a chemical industry zone: Removal, mass balance and source analysis. *Science China Chemistry* 55, 416–425.

Yunker, M.B., Macdonald, R.W., Vingarzan, R., Mitchell, R.H., Goyette D., Sylvestre, S. 2002. PAHs in the Fraser River Basin: A critical appraisal of PAHs ratios as indicators of PAH source and composition. *Organic Geochemistry* 33, 489–515.

Zogorski, J.S., Carter, J.M., Ivahnenko, T., Lapham, W.W., Moran, M.J., Rowe, B.L., Squillace, P.J., Toccalino, P.L. 2006. *Volatile organic compounds in the nation's ground water and drinking-water supply wells.* Circular 1292. U.S. Geological Survey.

6

Occurrence, Fate, and Removal of PAHs and VOCs in WWTPs Using Activated Sludge Processes and Membrane Bioreactors: Results from Italy and Greece

Evina Katsou,[1] **Simos Malamis,**[1] **Daniel Mamais,**[2] **David Bolzonella,**[1] **and Francesco Fatone**[1]

[1]*Department of Biotechnology, University of Verona, Verona, Italy*

[2]*Department of Water Resources and Environmental Engineering, School of Civil Engineering, National Technical University of Athens, Athens, Greece*

CONTENTS

Polycyclic aromatic hydrocarbons (PAHs) are aromatic hydrocarbons with two or more fused benzene rings originating from natural and anthropogenic sources. PAHs are widespread environmental contaminants in aquatic environments and mainly originate from fossil fuel combustion and from the release of petroleum and petroleum products (Tian et al. 2009, Jung et al. 2010, Timoney and Lee 2011). The PAHs have attracted environmental concern due to their ubiquitous occurrence, recalcitrance, bioaccumulation potential, and carcinogenic activity (Haritash and Kaushik 2009). Some PAHs with four or more benzene rings, such as benzo(a)anthracene, chrysene, and benzo(a)pyrene, can generate covalent DNA adducts and oxidative DNA lesions, resulting in mutagenic and carcinogenic effects (Cao et al. 2010, Arp et al. 2011). Moreover, PAHs are persistent in the environment and can cause long-term adverse effects (Liu et al. 2010, Patrolecco et al. 2010). PAH emissions in many countries have continuously increased owing to their rapid growth in energy consumption and the increase of traffic and shipping (Guo et al. 2010, Liu et al. 2010, Malik et al. 2011). Recently, accidents have released millions of barrels of oil containing massive amounts of PAHs into water bodies, such as in the Gulf of Mexico (Crone and Tolstoy 2010) and in the port of Dalian in China (Guo et al. 2009), causing significant damage to marine and freshwater ecosystems. As a result of the widespread use of PAHs and their adverse impacts on water ecosystems and human health, their removal from aquatic systems is of essence.

The occurrence, fate, and removal of PAHs from municipal wastewater treatment plants (WWTPs) has been investigated by several researchers (Manoli and Samara 1999, 2008; Blanchard et al. 2001; Busetti et al. 2006; Vogelsang et al. 2006; Fatone et al. 2011). PAHs are lipophilic (i.e., hydrophobic) chemicals, and the larger compounds are poorly water soluble and have lower volatility than smaller compounds. Several studies have reported that PAHs are sorbed on organic matter ranging from street particles to the waste-activated sludge produced by the WWTPs (Dobbs et al. 1989, Stringfellow and Alvarez-Cohen 1999). Directive 2013/39/EU of the European Parliament and of the Council: Directive Amending the Water Framework Directive 2000/60/EC and Directive 2008/105/EC (Directive 2013/39/EU 2013, Directive 2000/60/EC 2000, Directive 2008/105/EC 2008) regarding priority substances in the field of water policy has explicitly specified the PAHs anthracene, fluoranthene, and naphthalene in the list of priority substances in the field of water policy. Furthermore, the directive also specifies the general term *PAHs* in the priority pollutants list, which includes the compounds benzo(a)pyrene, benzo(b)fluoranthene, benzo(g,h,i)perylene, benzo(k)fluoranthene, and indeno(1,2,3-c,d)pyrene. Directive 2013/39/EU has set Environmental Quality Standards (EQS) for specific PAHs, which are summarized in Table 6.1.

Usually PAHs entering the WWTPs are notably adsorbed onto particulate organic matter due to their hydrophobic nature. Therefore, PAHs can be effectively removed during sedimentation of sewage sludge and are relocated from wastewater to sludge (Busetti et al. 2006). The adsorption of PAHs on dissolved and colloidal matter can improve PAH bioaccessibility in an anaerobic digester (Barret et al. 2010). Biotransformation/biodegradation is considered the major PAH degradation process (Haritash and Kaushik 2009) taking place in conventional activated sludge processes (CASPs) and in membrane bioreactors (MBRs) (Fatone et al. 2011). Other mechanisms, such as volatilization and chemical degradation, can

TABLE 6.1

Environmental Quality Standards (EQS) for PAHs and VOCs That Are Specified as Priority Substances within Directive 2013/39/EU

Priority Substance	Annual Average for Inland Surface Waters (µg/L)	Annual Average for Other Surface Waters (µg/L)	Maximum Allowable Concentration for Inland Surface Waters (µg/L)	Maximum Allowable Concentration for Other Surface Waters (µg/L)
Anthracene	0.1	0.1	0.1	0.1
Fluoranthene	0.0063	0.0063	0.12	0.12
Naphthalene	2	2	130	130
PAH	Not applicable	Not applicable	Not applicable	Not applicable
Benzene	10	8	50	50
1,2-Dichloroethane	10	10	Not applicable	Not applicable
Dichloromethane	20	20	Not applicable	Not applicable

contribute to the removal of PAHs, depending on the characteristics of the wastewater, process design, and system operating conditions (Manoli and Samara 1999, 2008; Byrns 2001; Cirja et al. 2008). The removal efficiencies and potential mechanisms of PAHs are well documented as the result of several studies. The reduction of the low molecular weight (LMW) PAHs is higher (33 to 99+%) than the high molecular weight (HMW) PAHs (18 to 60%) (Bergqvist et al. 2006). LMW PAHs were possibly removed due to biodegradation and volatilization, and HMW PAHs due to adsorption (Manoli and Samara 1999). Recent studies investigated the possible transformation from PAHs to oxygenated PAHs during the biological treatment process in WWTPs (Qiao et al. 2014).

High removal efficiencies of PAHs (92%) have been reported in CASPs (Sánchez-Ávila et al. 2009), while the level of removal in the chemical treatment is in the range of 61 to 78% (Vogelsang et al. 2006). Although the overall contribution of filtration to PAH removal has been proven small, membrane filtration can cause a decrease in the compound concentration of the MBR permeate. Filtration seems to have played a statistically significant role in the removal of anthracene in the cross-flow MBR, and of naphthalene and acenaphthene in the dead-end MBR (Mozo et al. 2011). This decrease can be related to the decrease in permeate chemical oxygen demand (COD) as a result of membrane filtration, but also to volatilization in the case of the dead-end system (aerated membranes).

Volatile organic compounds (VOCs) are a class of organic compounds with a vapor pressure greater than 0.1 mmHg at 20°C and 1 atm. These compounds are extensively used by several industries and, like PAHs, can adversely affect human health and the environment. VOCs are key ingredients in many consumer products, including fuels, paints, aerosols, cosmetics, disinfectants, refrigerants, and pesticides. It is therefore not surprising that they can find their way into municipal wastewater (Barceló 2004). The emissions of VOCs from WWTPs have been studied since the 1980s; aromatic VOCs typically account for more than 75% of the total VOC load (Namkung and Rittmann 1987). VOCs commonly met in WWTPs include benzene, trichloromethane, dichloromethane, toluene, tetrachloroethane, and trichloroethene (Ergas et al. 1995). The fate of aromatic VOCs during wastewater treatment is a major concern due to their volatilization and the resulting safety issues for the operators of the plant. Their bioaccumulation and biodegradation are also important factors that define the best approach to sewage sludge treatment and disposal. Despite the use of a variety of chemical controls on malodorous sewage streams, the release of sulfur compounds persists as a source of concern and complaint (Bianchi and Varney 1997). The European Union's Directive 2013/39/EU specifies the VOC compounds of benzene, 1,2-dichloroethane, and dichloromethane as priority pollutants, and the respective EQS are given in Table 6.1.

6.1 Fate and Removal of PAHs and VOCs in MBRs

Table 6.2 presents the results reported in the literature for several studies examining PAH occurrence and removal in the MBR and CASP processes for municipal and petrochemical wastewater.

As a general remark, the removal efficiencies obtained for petrochemical effluents are very high, owing to the high influent concentration and effective removal mechanisms in the bioreactors. In the case of municipal effluents, where the influent concentration was much lower, variable removals were obtained depending on the compound and on operating conditions.

Early results in the field of organic micropollutants removal by MBRs showed some superiority of MBR technology over CASP. This was attributed to the higher operating mixed liquor suspended solids (MLSS) concentration of MBRs and to the complete retention of suspended solids by MBRs, which could favor the development of slower-growing bacteria, and thus achieve better removal. However, recent research findings show that when MBR and CASP operate under similar operating conditions (i.e., hydraulic retention time (HRT), MLSS, solids retention time (SRT), temperature), the removal of organic micropollutants, including PAHs, by CASP and MBRs is similar (Joss et al. 2006, Weiss and Reemtsma 2008, Lesjean et al. 2011). Fatone et al. (2011) studied the occurrence, fate, and removal of 16 PAHs and 23 VOCs in five full-scale municipal CASPs, 3 of which also included MBRs operating in parallel with the CASP. The authors concluded that the removals of both VOCs and PAHs achieved by MBR and CASP were comparable. Most PAHs are mainly associated with the solid fraction of wastewater; the higher their solid-liquid

TABLE 6.2
PAHs ($\mu g\ L^{-1}$) in Influent and Effluent of MBR and CASP: Removal Efficiencies (%)

PAH	Stream	Influent ($\mu g\ L^{-1}$)	Effluent ($\mu g\ L^{-1}$)	Process	Removal (%)	Reference
Phenanthrene	Municipal	13.8	2.5	MBR		González-Pérez et al. (2012)
Fluoranthene		2.7	0.25			
Pyrene		2.8	0.23			
Benzene	Synthetic petrochemical	$\sim$8–9·10^5	$\sim$300–400	Dead-end MBR	99.95[a]/70.4[b]	Mozo et al. (2011)
Toluene		$\sim$8–9·10^5	$\sim$200		99.96[a]/55.3[b]	
Ethylbenzene		$\sim$5·10^5	$\sim$60–80		99.95[a]/92.9[b]	
Xylene		$\sim$5·10^5	$\sim$90		99.91[a]/60.1[b]	
Napthalene		$\sim$9·10^4	$\sim$100		99.99[a]/64.5[b]	
Acenaphthylene		$\sim$5–7·10^4	$\sim$600–700		97.20[a]/79.1[b]	
Acenapthrene		$\sim$10^4	$\sim$100		99.84[a]/61.7[b]	
Phenanthrene		$\sim$5·10^3	$\sim$4–5		99.20[a]/74.0[b]	
Anthracene		$\sim$10^3	$\sim$0.1		99.91[a]/95.9[b]	
Fluoranthene		$\sim$8–9·10^2	$\sim$0.1		99.68[a]/44.3[b]	
Benzene	Synthetic petrochemical	$\sim$8–9·10^5	$\sim$50	Dead-end MBR	99.62[a]/68.0[b]	Mozo et al. (2011)
Toluene		$\sim$8–9·10^5	$\sim$50		99.74[a]/63.1[b]	
Ethylbenzene		$\sim$5·10^5	$\sim$10–20		99.84[a]/88.9[b]	
Xylene		$\sim$5·10^5	$\sim$30–40		99.76[a]/75.6[b]	
Napthalene		$\sim$9·10^4	$\sim$1		98.63[a]/87.9[b]	
Acenaphthylene		$\sim$5–7·10^4	$\sim$100		90.61[a]/80.1[b]	
Acenapthrene		$\sim$10^4	$\sim$2–3		90.90[a]/70.2[b]	
Phenanthrene		$\sim$5·10^3	$\sim$3–4		99.16[a]/91.0[b]	
Anthracene		$\sim$10^3	$\sim$0.1		99.92[a]/98.5[b]	
Fluoranthene		$\sim$8–9·10^2	$\sim$0.3		99.83[a]/86.8[b]	

(Continued)

TABLE 6.2 (*Continued*)

PAHs (μg L^{-1}) in Influent and Effluent of MBR and CASP: Removal Efficiencies (%)

PAH	Stream	Influent (μg L^{-1})	Effluent (μg L^{-1})	Process	Removal (%)	Reference
Napthalene	Synthetic petrochemical	0.1, 4.0, 4.4, 12.0	0.1, 0.2, <LOQ, <LOQ	Submerged MBR	78	Wiszniowski et al. (2011)
Fluorene		8.2, 48.2, 62.2, 152.7	<LOQ, 0.3, 0.4, 0.2		>83	
Phenanthrene		43.2, 182.0, 322.2, 672.2	0.2, 0.4, 0.3, 0.3		>97	
Anthracene		5.4, 19.9, 35.4, 74.1	<LOQ, 0.2, 0.1, 0.1		89	
Naphthalene	Municipal	0.25	0.056	CASP 25,000 m^3 d^{-1}	31	Fatone et al. (2011)
Acenaphthylene		0.030	<0.005		>64	
Acenaphthene		0.18	<0.005		39	
Fluorene		0.177	0.019		32	
Phenanthrene		0.084	0.058		43	
Anthracene		0.014	<0.005		75	
Fluoranthene		0.028	0.017		>69	
Pyrene		0.025	0.017		>84	
Benzo(a)anthracene		0.023	0.013		69	
Chrysene		0.059	0.015			
Benzo(b)fluoranthene		0.016	<0.005			
Benzo(k)fluoranthene		0.032	<0.005			
Benzo(a)pyrene		0.016	<0.005			

(Continued)

TABLE 6.2 (Continued)

PAHs (µg L⁻¹) in Influent and Effluent of MBR and CASP: Removal Efficiencies (%)

PAH	Stream	Influent (µg L⁻¹)	Effluent (µg L⁻¹)	Process	Removal (%)	Reference
Indeno(1,2,3-c,d)pyrene		<0.005	<0.005		—	
Dibenz(a,h)anthracene		<0.005	<0.005		—	
Benzo(g,h,i)perylene		0.016	<0.005		>68.75	
Σ₁₆ PAHs		0.71	<0.195		≥73	
Naphthalene	Municipal	0.096	0.073[c] / 0.042[d]	CASP 15,000 m³ d⁻¹ / Pilot MBR operation in parallel with CASP	24[c] / 56[d]	Fatone et al. (2011)
Acenaphthylene		0.017	0.015[c] / <0.005[d]		12[c] / >71[d]	
Acenaphthene		0.084	0.025[c] / 0.014[d]		70[c] / 83[d]	
Fluorene		0.058	<0.005[c] / <0.005[d]		>91[c] / >91[d]	
Phenanthrene		0.052	0.014[c] / 0.012[d]		73[c] / 77[d]	
Anthracene		<0.005	<0.005[c] / 0.007[d]		—[c] / na[d]	
Fluoranthene		0.018	<0.005[c] / 0.048[d]		>72[c] / na[d]	
Pyrene		0.018	0.012[c] / 0.048[d]		33[c] / na[d]	
Benzo(a)anthracene		0.02	<0.005[c] / 0.007[d]		>75[c] / 65[d]	
Chrysene		<0.005	<0.005[c] / <0.005[d]		—[c] / —[d]	
Benzo(b)fluoranthene		<0.005	<0.005[c] / 0.00[d]		—[c] / —[d]	
Benzo(k)fluoranthene		<0.005	<0.005[c] / 0.00[d]		—[c] / —[d]	
Benzo(a)pyrene		<0.005	<0.005[c] / 0.00[d]		—[c] / —[d]	
Indeno(1,2,3–c,d)pyrene		<0.005	<0.005[c] / 0.00[d]		—[c] / —[d]	
Dibenz(a,h)anthracene		<0.005	<0.005[c] / 0.00[d]		—[c] / —[d]	
Benzo(g,h,i)perylene		<0.005	<0.005[c] / 0.00[d]		—[c] / —[d]	
Σ₁₆ PAHs		0.22	<0.139[c] / <0.178[d]		≥37[c] / ≥19[d]	

(Continued)

TABLE 6.2 (Continued)

PAHs (µg L⁻¹) in Influent and Effluent of MBR and CASP: Removal Efficiencies (%)

PAH	Stream	Influent (μg L^{-1})	Effluent (μg L^{-1})		Process	Removal (%)		Reference
Naphthalene	Municipal	0.113	0.043[c]	0.054[d]	CASP 15,000 m^3 d^{-1}	62[c]	52[d]	Fatone et al. (2011)
Acenaphthylene		<0.005	<0.005[c]	<0.005[d]	Full-scale MBR operation in parallel with CASP 4900 m^3 d^{-1}	—[c]	—[d]	
Acenaphthene		0.027	<0.005[c]	<0.005[d]		>81[c]	>81[d]	
Fluorene		0.008	<0.005[c]	<0.005[d]		>38[c]	>38[d]	
Phenanthrene		0.047	<0.005[c]	0.008[d]		>89[c]	83[d]	
Anthracene		<0.005	<0.005[c]	<0.005[d]		—[c]	—[d]	
Fluoranthene		<0.005	<0.005[c]	<0.005[d]		—[c]	—[d]	
Pyrene		<0.005	<0.005[c]	<0.005[d]		—[c]	—[d]	
Benzo(a)anthracene		0.008	0.007[c]	<0.005[d]		13[c]	>38[d]	
Chrysene		<0.005	<0.005[c]	<0.005[d]		—[c]	—[d]	
Benzo(b)fluoranthene		<0.005	<0.005[c]	<0.005[d]		—[c]	—[d]	
Benzo(k)fluoranthene		<0.005	<0.005[c]	<0.005[d]		—[c]	—[d]	
Benzo(a)pyrene		<0.005	<0.005[c]	<0.005[d]		—[c]	—[d]	
Indeno(1,2,3—c,d)pyrene		<0.005	<0.005[c]	<0.005[d]		—[c]	—[d]	
Dibenz(a,h)anthracene		<0.005	<0.005[c]	<0.005[d]		—[c]	—[d]	
Benzo(g,h,i)perylene		<0.005	<0.005[c]	<0.005[d]		—[c]	—[d]	
Σ_{16} PAHs		0.14	<0.05[c]	<0.062[d]		≥64[c]	≥56[d]	
Naphthalene	Municipal	0.634	0.074[c]	0.035[d]	CASP 118,000 m^3 d^{-1}	88[c]	94[d]	Fatone et al. (2011)
Acenaphthylene		0.022	0.011[c]	<0.005[d]	Pilot MBR operation in parallel with CASP (Italy)	50[c]	>77[d]	
Acenaphthene		0.285	0.023[c]	0.03[d]		92[c]	89[d]	

(Continued)

TABLE 6.2 (Continued)

PAHs ($\mu g\ L^{-1}$) in Influent and Effluent of MBR and CASP: Removal Efficiencies (%)

PAH	Stream	Influent ($\mu g\ L^{-1}$)	Effluent ($\mu g\ L^{-1}$)		Process	Removal (%)		Reference
Fluorene		0.148	0.011[c]	0.009[d]		93[c]	94[d]	
Phenanthrene		0.188	0.012[c]	0.022[d]		94[c]	88[d]	
Anthracene		0.038	<0.005[c]	<0.005[d]		>87[c]	>87[d]	
Fluoranthene		0.126	0.011[c]	0.017[d]		91[c]	87[d]	
Pyrene		0.107	0.015[c]	0.013[d]		86[c]	88[d]	
Benzo(a)anthracene		0.025	0.011[c]	<0.005[d]		56[c]	>80[d]	
Chrysene		0.025	<0.005[c]	<0.005[d]		>80[c]	>80[d]	
Benzo(b)fluoranthene		0.014	<0.005[c]	<0.005[d]		>64[c]	>64[d]	
Benzo(k)fluoranthene		<0.005	<0.005[c]	<0.005[d]		—[c]	—[d]	
Benzo(a)pyrene		0.014	<0.005[c]	<0.005[d]		>64[c]	>64[d]	
Indeno(1,2,3-c,d)pyrene		<0.005	<0.005[c]	<0.005[d]		—[c]	—[d]	
Dibenz(a,h)anthracene		<0.005	<0.005[c]	<0.005[d]		—[c]	—[d]	
Benzo(g,h,i)perylene		0.020	<0.005[c]	<0.005[d]		>75[c]	>75[d]	
Σ_{16} PAHs		1.54	<0.168[c]	<0.126[d]		≥89[c]	≥92[d]	
Naphthalene	Municipal	0.103	0.037		CASP 21,000 $m^3\ d^{-1}$ (Italy)	64		Fatone et al. (2011)
Acenaphthylene		0.011	<0.005			>55		
Acenaphthene		0.115	0.02			83		
Fluorene		0.043	0.016			63		
Phenanthrene		0.039	0.01			74		
Anthracene		0.013	<0.005			>62		
Fluoranthene		0.009	<0.005			>44		

(Continued)

TABLE 6.2 (Continued)

PAHs (µg L⁻¹) in Influent and Effluent of MBR and CASP: Removal Efficiencies (%)

PAH	Stream	Influent (µg L⁻¹)	Effluent (µg L⁻¹)	Process	Removal (%)	Reference
Pyrene		0.01	<0.005		>50	
Benzo(a)anthracene		0.01	<0.005		>50	
Chrysene		0.02	<0.005		>75	
Benzo(b)fluoranthene		0.02	<0.005		>75	
Benzo(k)fluoranthene		0.040	<0.005		>88	
Benzo(a)pyrene		0.020	<0.005		>75	
Indeno(1,2,3-c,d)pyrene		0.020	<0.005		>75	
Dibenz(a,h)anthracene		0.020	<0.005		>75	
Benzo(g,h,i)perylene		0.020	<0.005		>75	
Σ_{16} PAHs		0.32	<0.083		≥74	

PAH	Stream	Influent (µg L⁻¹)	Effluent (µg L⁻¹)	Process	Removal (%)			Reference
PAH	Petrochemical	>360		Full-scale MBR (Italy)	99%			Verlicchi et al. (2011)
Naphthalene	Municipal	7.3	5.0	CASP 40,000 m³ d⁻¹ (Greece)	—[e]	25.0[f]	12.3[g]	Manoli and Samara (2008)
Fluorene		0.7	0.23		28.6[e]	—[f]	54.0[g]	
Phenanthrene		1.7	0.20		66.5[e]	68.4[f]	—[g]	
Anthracene		0.09	0.007		63.3[e]	75.8[f]	12.5[g]	
Fluoranthene		0.24	0.029		70.0[e]	—[f]	90.0[g]	
Pyrene		0.47	0.06		74.5[e]	—[f]	91.4[g]	
Benzo(a)anthracene		0.05	0.005		70.0[e]	66.7[f]	—[g]	

(Continued)

TABLE 6.2 (Continued)

PAHs ($\mu g\ L^{-1}$) in Influent and Effluent of MBR and CASP: Removal Efficiencies (%)

PAH	Stream	Influent ($\mu g\ L^{-1}$)	Effluent ($\mu g\ L^{-1}$)	Process	Removal (%)		Reference
Chrysene		0.16	0.015		79.4[e]	57.6[f]	—[g]
B(e)Py		0.36	0.006		72.2[e]	40.0[f]	90.0[g]
Benzo(b)fluoranthene		0.023	0.007		—[c]	94.0[f]	—[g]
Benzo(k)fluoranthene		0.01	0.003		60.0[c]	25.0[f]	—[g]
Benzo(a)pyrene		0.22	0.005		95.5[e]	40.0[f]	16.7[g]
Dibenz(a,h)anthracene		0.05	0.002		96.0[e]	—[f]	—[g]
Benzo(g,h,i)perylene		0.029	0.008		65.5[e]	10.0[f]	11.1[g]
1,2,3-c,d indenopyrene		0.015	0.005		60.0[e]	—[f]	16.7[g]

Note: LOQ, limit of quantification.

[a] removed from liquid.

[b] degraded.

[c] the results obtained from the full-scale CASP.

[d] the results are obtained from the pilot or full-scale MBRs operated in parallel with the full-scale CASPs.

[e] %removal in primary treatment.

[f] %removal in secondary treatment.

[g] %removal in chlorination.

partitioning coefficient (i.e., K_P value), the higher their potential to sorb on particulate matter. K_P is expressed as a function of the octanol-water partitioning coefficient K_{OW}: $\log K_P = 0.58 \log K_{OW} + 1.14$ (Dobbs et al. 1989).

Several studies have been conducted to determine the main mechanisms involved in PAH removal from domestic wastewater by the MBR process. González-Pérez et al. (2012) developed mass balances on three selected PAHs (phenanthrene, fluoranthene, and pyrene) seen in an MBR having ultrafiltration membranes. The authors concluded that the MBR resulted in a high removal rate of the selected PAHs from urban wastewater, with performance levels in the range of 90%. The principal elimination route for PAHs in the MBR system was air stripping, which reduced the accumulation of PAHs in the biomass and limited the establishment of microorganisms capable of biodegrading or biotransforming these compounds. The contribution of biodegradation and sorption to the total compounds' removal was insignificant. In spite of the high hydrophobicity of the tested compounds, their accumulation in the biomass was very low and the sludge presented low PAH concentration. The concentration of four PAHs was reduced by more than 99% during the MBR operation for the treatment of petrochemical wastewater (Wiszniowski et al. 2011). The acclimation of activated sludge microorganisms to petroleum pollutants resulted in enhanced PAH biodegradation with time. The presence of surfactants in wastewater can increase the solubility of PAHs and potentially increase their bioavailability (Li and Chen 2009).

6.1.1 Critical Parameters That Affect PAH Removal

Several parameters can influence the fate and removal of micropollutants in the CASP and MBRs. These parameters are linked with the physicochemical properties of the compound (e.g., hydrophobicity, hydrophilicity, chemical structure), as well as with the operational parameters of the system (e.g., MLSS, SRT, pH, HRT, temperature, scale of the process, operation mode, etc.). A detailed analysis of the most influential parameters on the removal of PAHs in the MBR and CASP has been performed by Cirja et al. (2008). The operation mode and the scale of the process are critical factors, while data obtained from pilot, demonstration, or full-scale applications may significantly differ from those obtained from laboratory-scale experiments. However, the bench-scale examination can supply data on the fate of the target compounds, which may be extrapolated to a larger scale through modeling.

Hydrophobic compounds can be removed from the influent via sorption to the activated sludge. Compounds containing complex structures and toxic groups exhibit higher resistance to biodegradation (Cirja et al. 2008). As nonbiotic processes compete with biotic transformations, assessing and predicting the biodegradation rate of a specific aromatic compound is a difficult and important task. The SRT can be considered the most suitable design parameter to evaluate micropollutant removal in CASP and MBR (Byrns 2001, Lesjean et al. 2004, Clara et al. 2005, Joss 2005). Since micropollutant degradation is generally considered to increase with increasing SRT, MBRs were thought to have an advantage because they operate at higher SRT than CASP for a similar footprint. In MBRs, additional parameters that should be considered include the coarse bubble aeration that may enhance volatilization, hydrodynamic constraints due to liquid circulation, complete retention of suspended and colloidal matter, and accumulation of colloids within the mixed liquor. Thus, various amounts of dispersed bacteria and extracellular polymeric substances are generally observed in MBR sludge. These specifications vary with the MBR configuration and operation. The impact of SRT must be considered when evaluating the biodegradation of PAHs. The biodegradation of PAHs is usually enhanced when the SRT in the process is high enough (Cirja et al. 2008). Fatone et al. (2011) evaluated the effect of SRT on the removal of PAHs from municipal wastewater in an MBR where the SRT was very high, in the range of 200 to 500 days. The authors found a logarithmic relationship between the daily specific accumulation of PAHs in the activated sludge and the SRT. This suggests that increasing the SRT improved the removal of PAHs even in the range of very high SRT. However, the latter was partly offset by the lower biological activity, since higher SRT results in a lower volatile suspended solids/total suspended solids (VSS/TSS) ratio (i.e., lower biomass fraction) and lower maximum specific oxygen uptake rate (sOUR) and in situ sOUR (Fatone et al. 2011, Malamis et al. 2011). The MBR configuration also plays some role in the removal of PAHs (Mozo et al. 2011). Membranes can be submerged in the bioreactor or operated in an external element (side-stream configuration). In the latter case, the system can be operated with cross-filtration (high liquid velocity is then imposed at the membrane surface) or semi dead-end filtration with submerged membranes (coarse bubble aeration is then used to control fouling). Both systems generate shear stress, which reduces the size of the sludge flocs to a greater or lesser extent (Kim et al. 2001, Stricot et al. 2010). This phenomenon reduces the mass transfer resistance, which improves the accessibility of bacteria to pollutants and modifies the apparent biokinetic parameters (Fenu et al. 2010). However, the relation between sludge structure and biokinetics is complex and still controversial. Most PAH compounds become toxic at a given concentration. There is still no information on the long-term consequences of disaggregation in MBR, since acclimatization of the microbial population plays a key role in resistance development and biodegradation kinetics for xenobiotic compounds (Rezouga et al. 2009). The removal of hazardous aromatic compounds has been investigated in cross-flow and semi-dead-end MBRs. BTEX (benzene, toluene, ethylbenzene, xylene) and PAHs were efficiently eliminated (90 to 99.9%) from wastewater. However, the nonbiotic processes (i.e., volatilization, sorption) contributed significantly. The semi-dead-end MBR exhibited slightly higher removal of the studied organic micropollutants than the cross-flow MBR. In the dead-end configuration a thicker fouling layer is allowed to develop due to the lower shear. This layer may include a significant amount of biomass that contributes to organics degradation. It can also serve as a dynamic membrane, potentially enhancing the removal across the membrane. However, the nonbiotic processes were more significant in the dead-end MBR, while the degradation rates were higher in the cross-flow MBR due to the higher bioavailability of pollutants. The cross-flow filtration generates dispersed bacteria and larger quantities of dissolved and colloidal matter. Sorption of hydrophobic PAHs on suspended solids was lower in disaggregated sludge (Mozo et al. 2011). Wastewater temperature, which is subject to seasonal changes, influences the biodegradation rate, and thus the removal of PAHs. An increase in temperature of the order of 3 to 4°C can result in higher degradation of two-ring and three-ring PAHs. Manoli and Samara (1999) concluded that PAHs having lower molecular weight (e.g., naphthalene, acenaphtene, phenanthrene, anthracene, fluorene) undergo substantial reduction (>40%) in biological systems due to volatilization and biodegradation.

The removal of VOCs from municipal wastewater with the use of MBR has received limited attention. Table 6.3 summarizes the results of works reported in literature for VOC removal from wastewater employing the MBR and the CASP process. Min and Ergas (2006) examined the volatilization and biodegradation rates for acetaldehyde, butyraldehyde, and vinyl acetate in an MBR for varying organic loading rates (OLRs) and found that the overall removal of the three VOCs was higher than 99.7%. Biodegradation and volatilization are the main competing VOC removal mechanisms; their contribution to the overall removal of VOCS depends on several operating parameters, including OLR, SRT, and dissolved oxygen (DO) concentration. A biological reactor that is designed at long SRT and low DO can enhance the biodegradation and minimize the volatilization removal rate. MBR has been proven as a promising solution to reduce VOC emissions from wastewater (Min and Ergas 2006). A pilot-scale MBR operating in Athens (Greece) was able to effectively (>85%) remove dichloromethane, trichloroethene, toluene, octane, nonane, ethylbenzene, d-limonene, p-xylene, o-xylene, nonane, alphapinene, dimethyl disulfide, dimethyl trisulfide, and decane (Table 6.3). The MBR accomplished poor removal of hexane; its removal is limited by its slow biodegradation, which is attributed to its low bioavailability (Lee et al. 2010, Yang et al. 2010). The treated secondary effluent in municipal WWTPs is in essence free of most of the VOCs. The concentration of several VOCs has been reported to be below, or close to, the limit of quantification. The removal of acetone was found to be low in both MBR and CASP (Table 6.3). On the contrary, dichloromethane, which is a poorly biodegradable compound, was more effectively removed than acetone (i.e., >84.8% by the MBR and 64.3% by the CASP). The latter shows that other removal mechanisms apart from biodegradation seem to be important, in particular volatilization. Fallah et al. (2010) found that styrene removal from the liquid phase by MBR was consistently higher than 99%. The MBR achieved very high removal efficiencies with negligible volatilization of styrene in the exit air stream. Both MBR and CASP were effective in the removal of biogenic compounds (dimethyl disulfide and dimethyl trisulfide) from wastewater (Table 6.3).

TABLE 6.3

VOCs (μg L⁻¹) in Influent and Effluent of MBR and CAS Plants: Removal Efficiencies (%)

VOC	Wastewater Stream	Influent μg L⁻¹ (mean)	Effluent μg L⁻¹ (mean)	Process	Removal (%)	Reference
Benzene	Municipal	0.21	0.06	CASP, 25,000 m³ day⁻¹	71	Fatone et al. (2011)
Toluene		3.544	<0.005		—	
Ethylbenzene		0.238	0.03		87	
m-Xylene + p-xylene		0.568	0.04		93	
o-Xylene		0.035	0.02		43	
Styrene		0.148	0.005		97	
1,2,4-Trimethylbenzene		0.168	0.02		88	
4-Chlorotoluene		0.05	<0.005		>90	
Σ BTEXS			0.16		≥82	
Benzene	Municipal	<0.005	<0.005[a] <0.005[b]	CASP, 15,000 m³ day⁻¹	—[a] —[b]	Fatone et al. (2011)
Toluene		4.885	2.7[a] 0.52[b]	Pilot MBR operation in parallel with CASP	44[a] 89[b]	
Ethylbenzene		0.093	<0.005 0.12[b]		>95[a] na[b]	
m-Xylene + p-xylene		0.15	0.08[a] 0.35[b]		47[a] na[b]	
o-Xylene		0.192	0.19[a] 0.2[b]		1[a] na[b]	
Styrene		0.139	0.28[a] 0.2[b]		na[a] na[b]	
1,2,4-Trimethylbenzene		1.018	0.71[a] 0.34[b]		30[a] 67[b]	
4-Chlorotoluene		0.301	<0.005[a] 0.27[b]		>98[a] 10[b]	
Σ BTEXS			3.26[a] 1.40[b]		≥37[a] ≥45[b]	

(Continued)

TABLE 6.3 (*Continued*)

VOCs (µg L^{-1}) in Influent and Effluent of MBR and CAS Plants: Removal Efficiencies (%)

VOC	Wastewater Stream	Influent µg L^{-1} (mean)	Effluent µg L^{-1} (mean)		Process	Removal (%)		Reference
Benzene	Municipal	0.063	<0.005[a]	<0.005[b]	CASP, 15,000 m³ day⁻¹	>92[a]	>92[b]	Fatone et al. (2011)
Toluene		3.008	0.84[a]	0.7[b]	Full-scale MBR operation in parallel with CASP, 4900 m³ day⁻¹	72[a]	77[b]	
Ethylbenzene		0.132	0.13[a]	<0.005[b]		2[a]	>96[b]	
m-Xylene + p-xylene		0.193	<0.005[a]	<0.005[b]		>97[a]	>97[b]	
o-Xylene		0.313	0.17[a]	0.1[b]		46[a]	68[b]	
Styrene		0.226	0.005[a]	0.1[b]		98[a]	56[b]	
1,2,4-Trimethylbenzene		0.764	<0.005[a]	0.5[b]		>99[a]	n[ab]	
4-Chlorotoluene		0.306	<0.005[a]	<0.005[b]		>98[a]	>98[b]	
Σ BTEXS			1.16[a]	0.92[b]		≥68[a]	≥81[b]	
Benzene	Municipal	0.239	<0.005[a]	<0.005[b]	CASP, 118,000 m³ day⁻¹	>98[a]	>98[b]	Fatone et al. (2011)
Toluene		7.169	2.7[a]	1.9[b]	Pilot MBR operation in parallel with CASP	62[a]	73[b]	
Ethylbenzene		0.216	0.12[a]	<0.005[b]		44[a]	>98[b]	
m-Xylene + p-xylene		0.775	0.45[a]	0.25[b]		42[a]	68[b]	
o-Xylene		0.351	0.3[a]	0.35[b]		15[a]	n[ab]	
Styrene		1.116	0.16[a]	0.005[b]		100[a]	100[b]	
1,2,4-Trimethylbenzene		0.81	0.41[a]	0.32[b]		49[a]	n[ab]	
4-Chlorotoluene		0.22	0.1[a]	<0.005[b]		55[a]	>98[b]	
Σ BTEXS			3.74[a]	2.52[b]		≥60[a]	≥73[b]	
Benzene	Municipal	0.063	<0.005		CASP, 21,000 m³ day⁻¹ (Italy)			Fatone et al. (2011)
Toluene		3.008	0.75			75		
Ethylbenzene		0.093	<0.005			>95		

(*Continued*)

TABLE 6.3 (Continued)

VOCs ($\mu g\ L^{-1}$) in Influent and Effluent of MBR and CAS Plants: Removal Efficiencies (%)

VOC	Wastewater Stream	Influent $\mu g\ L^{-1}$ (mean)		Effluent $\mu g\ L^{-1}$ (mean)		Process	Removal (%)		Reference
m-Xylene + p-xylene		0.15		<0.005			>97		
o-Xylene		0.035		<0.005			>86		
Styrene		0.139		0.005			96		
1,2,4-Trimethylbenzene		0.168		<0.005			>97		
4-Chlorotoluene		0.05		<0.005			>90		
Σ BTEXS				0.78			≥90		
Acetaldehyde	Municipal					MBR, OLR = 1.1, kgCOD $m^{-3}\ day^{-1}$	99.05[c]	0.80[d]	Min and Ergas (2006)
Butyraldehyde							97.72[c]	2.06[d]	
Vinyl acetate							93.87[c]	6.09[d]	
Acetaldehyde	Municipal					MBR, OLR = 1.5, kgCOD $m^{-3}\ day^{-1}$	88.38[c]	11.47[d]	Min and Ergas (2006)
Butyraldehyde							98.03[c]	1.72[d]	
Vinyl acetate							87.43[c]	12.50[d]	
Acetaldehyde	Municipal					MBR, OLR = 1.5, kgCOD $m^{-3}\ day^{-1}$	98.98[c]	0.87[d]	Min and Ergas (2006)
Butyraldehyde							97.56[c]	2.19[d]	
Vinyl acetate							99.40[c]	0.59[d]	
Dichloromethane	Municipal	4.646[a]	3.650[b]	1.659[a]	0.553[b]	CASP, 20,000 $m^{-3}\ day^{-1}$, (Greece)	64.3[a]	84.8[b]	NTUA, unpublished data
Hexane		0.412[a]	0.312[b]	0.251[a]	0.130[b]	Pilot-scale MBR, 0.5 $m^{-3}\ day^{-1}$	39.1[a]	58.2[b]	

(Continued)

TABLE 6.3 (*Continued*)

VOCs (μg L⁻¹) in Influent and Effluent of MBR and CAS Plants: Removal Efficiencies (%)

VOC	Wastewater Stream	Influent μg L⁻¹ (mean)		Effluent μg L⁻¹ (mean)		Process	Removal (%)		Reference
Trichloromethane		14.524[a]	17.619[b]	3.889[a]	3.016[b]		73.2[a]	82.9[b]	
Benzene		0.658[a]	0.110[b]	0.110[a]	0.110[b]		83.3[a]	—[b]	
Trichloroethene		4.005[a]	4.526[b]	0.403[a]	0.059[b]		89.9[a]	98.7[b]	
Toluene		47.684[a]	28.824[b]	0.772[a]	0.441[b]		98.4[a]	98.5[b]	
Octane		0.352[a]	0.314[b]	0.200[a]	0.020[b]		43.2[a]	93.6[b]	
Ethylbenzene		0.628[a]	0.293[b]	0.251[a]	0.105[b]		60.0[a]	64.3[b]	
p-Xylene		0.764[a]	0.350[b]	0.382[a]	0.080[b]		50.0[a]	77.3[b]	
Nonane		0.497[a]	0.393[b]	0.012[a]	0.014[b]		97.7[a]	96.3[b]	
Styrene		0.885[a]	0.221[b]	0.221[a]	0.221[b]		75.0[a]	—[b]	
o-Xylene		1.596[a]	0.610[b]	0.563[a]	0.117[b]		64.7[a]	80.8[b]	
Alpha-pinene		0.218[a]	0.151[b]	0.019[a]	0.003[b]		91.4[a]	97.8[b]	
Decane		0.013[a]	0.006[b]	0.008[a]	0.004[b]		37.5[a]	43.6[b]	
1,2,4-Trimethylbenzene		0.415[a]	0.115[b]	0.115[a]	0.115[b]		72.2[a]	—[b]	
d-Limonene		0.854[a]	0.524[b]	0.854[a]	0.146[b]		—[a]	72.1[b]	
1,2,3-Trimethylbenzene		0.333[a]	0.104[b]	0.104[a]	0.104[b]		68.8[a]	—[b]	
Dimethyl disulfide		55.660[a]	49.057[b]	6.918[a]	0.786[b]		87.6[a]	98.4[b]	
Dimethyl trisulfide		4.824[a]	3.259[b]	0.913[a]	0.326[b]		81.1[a]	90.0[b]	

[a] The results obtained from the full-scale CASP.
[b] The results are obtained from the pilot- or full-scale MBRs operated in parallel with the full-scale CASPs.
[c] Biodegradation.
[d] Volatilization.

6.2 Results from Selected Case Studies in Italy and Greece

6.2.1 Italian Case Studies

The current Italian law (decree 152/06) states that liquid waste and industrial wastewater may both be collected by municipal WWTPs if the total hydrocarbon and aromatic organic solvent concentrations are lower than 10 and 0.4 mg L^{-1}, respectively. In Italy there are several large petrochemical and refinery sites producing large quantities of industrial effluents that contain fuel-derived aromatic hydrocarbons. These compounds can therefore find their way to municipal wastewater. Fatone et al. (2011) investigated the occurrence, fate, and removal of 16 PAHs recommended by the USEPA and 23 VOCs, including 17 aromatic VOCs and BTEX in five full-scale Italian WWTPs (having design capacities of 12,000 to 700,000 population equivalent), as well as 3 MBR facilities (one full-scale and two pilot-scale) operating in the premises of the examined CASP. The CASP performance was compared against that of MBR, in order to provide some insight as to when MBR technology results in enhanced removal of PAHs and VOCs.

6.2.1.1 Occurrence, Removal, and Fate of Aromatic VOCs

In the influent (raw) wastewater of the five Italian municipal WWTPs, toluene, ethylbenzene, xylene, and styrene were the most abundant VOCs, together with 1,2,4-trimethylbenzene and 4-chlorotoluene (Table 6.3). The removal of aromatic VOCs by CASP and MBR was almost complete, and the secondary effluents often showed VOC concentrations below the limits of quantification (Table 6.3). Fatone et al. (2011) reported that from the 23 studied VOCs, only toluene was consistently present in secondary effluents at appreciable levels, in the ranges of <0.005 to 2.7 µg L^{-1} for CASP and 0.5 to 1.9 µg L^{-1} for MBR. The authors attributed this to its high influent concentration in combination with its low bioavailability.

The relative VOC removal efficiencies varied among the different sections of the WWTPs. The headworks and primary treatment removed 40 to 60% of the VOCs from the raw wastewater, while the subsequent biological treatment stage removed 10 to 50% of the remaining VOCs (after the removal by the headworks and primary treatment). The removal of VOCs within the headworks is an issue that should be dealt with, since most WWTPs do not implement off-gassing control strategies. Specifically, the turbulence in the headworks, as well as air stripping if an aerated grit chamber is used, enhances VOC removal through volatilization in the pretreatment stage (Metcalf and Eddy 2003).

Despite the fact that different treatment technologies are implemented in the headworks and the main wastewater treatment line, this did not affect the VOC concentrations in the secondary effluent. Consequently, the aromatic VOC content in the effluents was not clearly dependent on CASP or

MBR technology. The authors found that sorption was not involved in the removal of aromatic VOCs, since their concentration in sewage sludge was below the detection limit (0.5 mg/kg TS). Volatilization was probably the major removal mechanism for aromatic VOCs. The MBR technology did not provide any significant advantage in terms of VOC removal despite the fact that coarse bubble aeration, which is employed to reduce membrane fouling, may enhance volatilization within the biological reactor.

6.2.1.2 Occurrence, Fate, and Removal of PAHs

The concentration of 16 PAHs in the raw municipal influent of the five Italian WWTPs was lower than 1.0 µg L^{-1}, and the total values were all less than 2 µg L^{-1}, even for large WWTPs, where most of the surrounding catchment area is urbanized (Table 6.2). In the raw sewage samples, naphthalene was detected at a frequency of 95% with an average concentration of 0.24 µg L^{-1}. Phenanthrene, fluorene, and acenaphthene were detected at a frequency of 85, 71, and 56%, respectively, above the limit of detection in the raw urban wastewaters. All the other PAHs examined had a frequency of occurrence below 50%. The concentration of any single compound except naphthalene was found to be lower than 0.3 µg L^{-1}. PAHs were mainly associated ($\geq$65%) with particulate matter in raw sewage. This finding is in agreement with other works that show that PAHs are mainly associated with particulate matter in sewage (Lau and Stenstrom 2005, Mansuy-Huault et al. 2009). The headworks and primary treatment stages removed PAHs by 40 to 60%, while 10 to 90% of the remaining PAHs were removed by the subsequent biological processes. These results were in line with the studies performed in other full-scale WWTPs (Manoli and Samara 2008), in which PAH removal ranged from 28 to 67% in the primary treatment, including primary sedimentation. On the other hand, naphthalene was almost completely volatilized in the headworks due to turbulence. The apparent removal levels of PAHs in CASP plants and MBRs were comparable, but the actual removals, which were related to biodegradation of the PAHs, were enhanced by the long SRT and showed a logarithmic relationship with the sludge age.

6.2.2 The Greek Case Studies

6.2.2.1 Occurrence, Fate, and Removal of Aromatic VOCs

In influent raw wastewater samples from three of the largest municipal CASPs—Metamorphosis, Psyttalia, and Volos—in Greece, annual mean concentrations of 40 VOCs ranged from 0.06 µg L^{-1} for xylene to 162 µg L^{-1} for 1,1,1-trichloroethane (Nikolaou et al. 2002). At the Metamorphosis and Volos WWTPs, only trichloroethene was detected in secondary effluent samples. At Psyttalia WWTP, which at the time did not include a biological treatment stage, most VOCs were also detected in the primary effluent, but at significantly lower concentrations. Nikolaou et al. (2002) reported average concentrations of toluene, tetrachloroethylene, and dibromochloromethane in primary effluent samples from Psyttalia WWTP equal to 6.83, 5.50, and 3.33 µg L^{-1}, respectively.

The removal of VOCs from municipal wastewater was studied using an immersed MBR compared to a CASP. The MBR was installed at the premises of a municipal WWTP and received the same influent as the full-scale CASP. Thirty-five VOCs were qualitatively determined; of these, 19 were further selected for quantitative determination. The MBR achieved effective and relatively higher removal of VOCs than CASP (Table 6.3). Specifically, for 15 out of the 19 VOCs that were quantitatively determined, the MBR permeate had a concentration lower than 0.3 µg L^{-1}, while for the CASP-treated effluent, 8 compounds were below this value. The MBR was able to achieve high removal efficiencies for the volatile sulfide compounds (VSCs) dimethyl disulfide and dimethyl trisulfide (Table 6.3). The high VOC removal by the MBR process was mainly attributed to the enhanced volatilization due to coarse bubble aeration. Biodegradation may not be the dominant removal mechanism, since compounds that are considered to be easily biodegradable (e.g., acetone) were less effectively removed than other compounds that were poorly biodegradable (e.g., dichloromethane). The removal of VOCs through volatilization was considered to be important in the MBR process due to coarse bubble aeration or the compactness of the system. Adsorption on suspended solids may also contribute to VOC removal. Several hydrophobic VOCs such as hexane, p-xylene, o-xylene, and toluene were not effectively removed by the CASP (NTUA).

6.2.2.2 Occurrence, Removal, and Fate of Aromatic PAHs

The removal of PAHs was investigated by Manoli and Samara (2008) in the wastewater treatment process in the CASP WWTP of Thessaloniki in northern Greece. Samples were collected from various stages of a CAS treatment facility. The 16 PAHs recommended by the EPA were found at quantifiable concentrations in the raw wastewater, ranging from a low median of 4.0 ng L^{-1} to a maximum median value of 2800 ng L^{-1} for naphthalene (Table 6.2) (Manoli and Samara 2008). The frequency of appearance of all PAHs, except acetone, was 100%. Besides naphthalene, all median concentrations of each individual PAH were below 1 µg L^{-1}. Between the various wastewater treatment stages, higher removal efficiencies were obtained in primary treatment in the range of 28 to 67%. Higher removal was obtained for the more hydrophobic compounds. PAH concentration in the primary sludge ranged from 36.6 µg kg^{-1} dry solids for benzofluoranthene to 4645 µg kg^{-1} dry solids for phenanthrene (Samara et al. 1995). The sum of the six PAHs compounds specified by the World Health Organization (WHO), namely, fluoranthene, benzo(b)fluoranthene, benzo(k)fluoranthene, benzo(a)pyrene, benzo(g,h,i) perylene, and indeno(1,2,3-c,d)pyrene, was equal to 1052 µg kg^{-1} dry solids. It should be emphasized that these values are significantly lower than any limits that have been proposed for sludge use on land (Directive 2000/60/EC 2000).

In the biological stage, PAH removal efficiency ranged between <1 and 61%, and in the whole process, from 37 to 89%. Chlorination did not appear to significantly affect the removal of PAHs, yielding mean removals between <1 and 10% for all PAH compounds. Manoli and Samara (2008) reported that of the 16 PAHs studied, only fluorene and naphthalene were determined in the nonchlorinated secondary effluent samples at appreciable median concentrations, equal to 0.50 and 0.33 µg L^{-1}, respectively.

6.3 CASP vs. MBR (Overall Evaluation)

MBR technology has been successfully applied for treatment of industrial wastewater at the source and for centralized and decentralized treatment of municipal sewage (Fane and Fane 2005, Judd 2006, Lesjean et al. 2008, Di Fabio et al. 2013, Malamis et al. 2014). The MBR process is accepted as a technology of choice for wastewater treatment, and the market is still showing sustained growth. One of the early misconceptions related to MBR is its potential to achieve higher removal of organic micropollutants, including VOCs and PAHs (Lesjean et al. 2011). A number of research works have reported that the effectiveness of MBR technology in the removal of xenobiotics and persistent compounds is not sufficiently pronounced to serve as the sole justification for employing MBRs in municipal wastewater treatment (DeWever et al. 2007, Weiss and Reemtsma 2008).

Concerning the concentration of PAHs in the water discharged from an MBR, it is reasonable to assume that the MBR provides an advantage by retaining the colloidal nonsettling particulates containing adsorbed PAHs, which would pass through a conventional settling tank. Recent evidence suggests that this fact is not important to the overall MBR operation. Fatone et al. (2011) showed that the PAH concentrations in the MBR permeate are similar to that of the secondary effluent of a CASP that operates in parallel with the MBR. Since residual PAHs were dissolved in the secondary effluent, there would be little advantage to incorporating the membrane systems used in MBRs. The full-scale and pilot MBR systems operating in parallel resulted in removal levels analogous to those of the CASP (80 to 95%).

Based on the results of the Greek case study, the CASP exhibited similar or lower removal efficiency of most VOCs than the MBR. The intense and continuous coarse bubble aeration probably enhanced the removal of certain VOCs via volatilization compared to the CASP. In this pilot MBR, only an aerobic chamber existed in which the membrane unit was submerged. Therefore, air scouring was supplied to the whole of the bioreactor. However, other studies reported no major influences of volatilization within the aerated filtration chamber (Fatone et al. 2011), as this was located at the end of the aerobic reactor where the major volatilization processes take place. The authors concluded that aromatic VOCs had comparable removal efficiencies in CASP and MBR processes since volatilization and air stripping were most probably the main removal mechanisms. Moreover, even VOCs could be influenced in different ways by MBRs in which strong coarse bubble aeration is used to scour the submerged membranes.

6.4 Conclusions

Naphthalene, phenanthrene, fluorene, and acenaphthene are PAHs that frequently occur in municipal WWTPs. Benzene, trichloromethane, dichloromethane, toluene, tetrachloroethane, and trichloroethene are VOCs often occurring in municipal effluents. Recent literature shows that MBR and CASPs result in similar removal of VOCs and PAHs from wastewater. PAHs are mainly sorbed to the activated sludge, while volatilization seems to be the main removal mechanism for VOCs. The pretreatment and primary treatment processes contribute significantly (40 to 60%) to the removal of VOCs and PAHs. The subsequent biological treatment stage usually removes 10 to 50% of the remaining VOCs and 10 to 90% of the remaining PAHs.

Acknowledgments

The authors thank the Bodossaki Foundation for financially supporting this work. The Italian Ministry of University and Research is kindly acknowledged for the funding to projects PRIN2003 and PRIN 2005.

References

Arp, H.P.H., Azzolina, N.A., Cornelissen, G., Hawthorne, S.B. 2011. Predicting pore water EPA-34 PAH concentrations and toxicity in pyrogenic-impacted sediments using pyrene content. *Environ. Sci. Technol.* 45; 5139–5146.

Barceló, D., ed. 2004. *Emerging organic pollutants in waste waters and sludge*. London: Springer.

Barret, M., Carrere, H., Delgadillo, L., Patureau, D. 2010. PAH fate during the anaerobic digestion of contaminated sludge: Do bioavailability and/or cometabolism limit their biodegradation? *Water Res.* 44; 3797–3806.

Bergqvist, P.A., Augulyt, L., Jurjonien, V. 2006. PAH and PCB removal efficiencies in Umea (Sweden) and Siauliai (Lithuania) municipal wastewater treatment plants. *Water Air Soil Pollut.* 175; 291–303.

Bianchi, A.P., Varney, M.S. 1997. Volatilization processes in wastewater treatment plants as a source of potential exposure to VOCs. *Ann. Occup. Hyg.* 41; 437–454.

Blanchard, M., Teil, M.J., Ollivon, D., Garban, B., Chestérikoff, C.C., Chevreuil, M. 2001. Origin and distribution of polyaromatic hydrocarbons and polychlorobiphenyls in urban effluents to wastewater treatment plants of the Paris area (France). *Water Res.* 35; 3679–3687.

Busetti, F., Heitz, A., Cuomo, M., Badoer, S., Traverso, P. 2006. Determination of sixteen polycyclic aromatic hydrocarbons in aqueous and solid samples from an Italian wastewater treatment plant. *Chromatogr. A*, 1102; 104–115.

Byrns, G. 2001. The fate of xenobiotic organic compounds in wastewater treatment plants. *Water Res.* 35; 2523–2533.

Cao, Z., Liu, J., Luan, Y., Li, Y., Ma, M., Xu, J., Han, S. 2010. Distribution and ecosystem risk assessment of polycyclic aromatic hydrocarbons in the Luan River, China. *Ecotoxicology* 19; 827–837.

Cirja, M., Ivashechkin, P., Schaffer, A., Corvini, P.F.X. 2008. Factors affecting the removal of organic micropollutants from wastewater in conventional treatment plants (CTP) and membrane bioreactors (MBR). *Rev. Environ. Sci. Biotechnol.* 7; 61–78.

Clara, M., Kreuzinger, N., Strenn, B., Gans, O., Kroiss, H. 2005. The solids retention time—A suitable design parameter to evaluate the capacity of wastewater treatment plants to remove micropollutants. *Water Res.* 39; 97–106.

Crone, T.J., Tolstoy, M. 2010. Magnitude of the 2010 Gulf of Mexico oil leak. *Sci. Total Environ.* 330; 634.

DeWever, H., Weiss, S., Reemtsma, T., Vereecken, J., Muller, J., Knepper, T., Rorden, O., Gonzalez, S., Barceló, D., Hernando, M.D. 2007. Comparison of sulfonated and other micropollutants removal in membrane bioreactor and conventional wastewater treatment. *Water Res.* 41; 935–945.

Di Fabio, S., Malamis, S., Katsou, E., Vecchiato, G., Cecchi, F., Fatone, F. 2013. Are centralized MBRs coping with the current transition of large petrochemical areas? A pilot study in Porto-Marghera (Venice). *Chem. Eng. J.* 214; 68–77.

Directive 2000/60/EC. 2000. Directive 2000/60/EC of the European Parliament and of the Council of 23 October 2000 establishing a framework for community action in the field of water policy. *Official Journal of the European Union* L 327.1-73.

Directive 2008/105/EC. 2008. Directive 2008/105/EC of the European Parliament and of the Council of 16 December 2008 on environmental quality standards in the field of water policy. *Official Journal of the European Union* L 348/84.

Directive 2013/39/EU. 2013. Directive 2013/39/EU of the European Parliament and of the Council of 12 August 2013 amending Directives 2000/60/EC and 2008/105/EC as regards priority substances in the field of water policy. *Official Journal of the European Union* L 226/1.

Dobbs, R.A., Wang, L., Govind, R. 1989. Sorption of toxic organic compounds on wastewater solids: Correlation with fundamental properties. *Environ. Sci. Technol.* 23; 1092–1097.

Ergas, S.J., Schroeder, E.D., Chang, D.P.Y., Morton, R.L. 1995. Control of volatile organic compound emissions using a compost biofilter. *Water Environ. Res.* 67; 816–821.

Fallah, N., Bonakdarpour, B., Nasernejad, B., AlaviMoghadam, M.R. 2010. Long-term operation of submerged membrane bioreactor (MBR) for the treatment of synthetic wastewater containing styrene as volatile organic compound (VOC): Effect of hydraulic retention time (HRT). *J. Hazard. Mater.* 178; 718–724.

Fane, A.G., Fane, S.A. 2005. The role of membrane technology in sustainable decentralized wastewater systems. *Water Sci. Technol.* 51; 317–325.

Fatone, F., Di Fabio, S., Bolzonella, D., Cecchi, F. 2011. Fate of aromatic hydrocarbons in Italian municipal wastewater systems: An overview of wastewater treatment using conventional activated-sludge processes (CASP) and membrane bioreactors (MBRs). *Water Res.* 45; 93–104.

Fenu, A., Guglielmi, G., Jimenez, J., Sperandio, M., Saroj, D., Lesjean, B., Brepols, C., Thoeye, C., Nopens, I. 2010. Activated sludge model (ASM) based modelling of membrane bioreactor (MBR) processes: A critical review with special regard to MBR specificities. *Water Res.* 44; 4272–4294.

González-Pérez, D.M., Garralón, G., Plaza, F., Pérez, J.I., Moreno, B., Gómez, M.A. 2012. Removal of low concentrations of phenanthrene, fluoranthene and pyrene from urban wastewater by membrane bioreactors technology. *J. Environ. Sci. Health A* 47; 2190–2197.

Guo, W., Wang, Y., Me, M., Cui, Y. 2009. Modeling oil spill trajectory in coastal waters based on fractional Brownian motion. *Mar. Pollut. Bull.* 58; 1339–1346.

Guo, J., Wu, F., Luo, X., Liang, Z., Liao, H., Zhang, R., Li, W., Zhao, X., Chen, S., Mai, B. 2010. Anthropogenic input of polycyclic aromatic hydrocarbons into five lakes in Western China. *Environ. Pollut.* 158; 2175–2180.

Haritash, A.K., Kaushik, C.P. 2009. Biodegradation aspects of polycyclic aromatic hydrocarbons (PAHs): A review. *J. Hazard. Mater.* 169; 1–15.

Joss, A. 2005. Removal of pharmaceuticals and fragrances in biological wastewater treatment. *Water Res.* 39; 3139–3152.

Joss, A., Zabczynski, S., Göbel, A., Hoffmann, B., Löffler, D., McArdell, C.S., Ternes, T.A., Thomsen, A., Siegrist, H. 2006. Biological degradation of pharmaceuticals in municipal wastewater treatment: Proposing a classification scheme. *Water Res.* 40; 1686–1696.

Judd, S. 2006. *The MBR book: Principles and applications of membrane bioreactors in water and wastewater treatment.* Oxford: Elsevier.

Jung, J.E., Lee, D.S., Kim, S.J., Kim, D.W., Kim, S.K., Kim, J.G. 2010. Proximity of field distribution of polycyclic aromatic hydrocarbons to chemical equilibria among air, water, soil, and sediment and its implications to the coherence criteria of environmental quality objectives. *Environ. Sci. Technol.* 44; 8056–8061.

Kim, J.S., Lee, C.H., Chang, I.S. 2001. Effect of pump shear on the performance of a cross flow membrane bioreactor. *Water Res.* 35; 2137–2144.

Lau, S.L., Stenstrom, M.K. 2005. Metals and PAHs adsorbed to street particles. *Water Res.* 39; 4083–4092.

Lee, E.H., Kim, J., Cho, K.S., Ahn, Y.G., Hwang, G.S. 2010. Degradation of hexane and other recalcitrant hydrocarbons by a novel isolate, *Rhodococcus* sp. EH831. *Environ. Sci. Pollut. Res.* 17; 64–77.

Lesjean, B., Gnirss, R., Buisson, H., Keller, S., Tazi-Pain, A., Luck, F. 2005. *Outcomes of a 2-year investigation on enhanced biological nutrients removal and trace organics elimination in membrane bioreactor (MBRs).* Water Science & Technology, 52(10–11); 453–460.

Lesjean, B., Gnirss, R., Vocks, M., Luedicke, C. 2008. In *Does MBR represent a viable technology for advanced nutrients removal in wastewater treatment of small communities?* EWA/JSWA/WEF—3rd Joint Specialty Conference "Sustainable Water Management in Response to 21st Century Pressures," IFAT Munich, Germany, May 5–9.

Lesjean, B., Tazi-Pain, A., Thaure, D., Moeslang, H., Buisson, H. 2011. Ten persistent myths and the realities of membrane bioreactor technology for municipal applications. *Water Sci. Technol.* 63; 32–39.

Li, J.-L., Chen, B.-H. 2009. Surfactant-mediated biodegradation of polycyclic aromatic hydrocarbons. *Materials* 2; 76–94.

Liu, C., Tian, F., Chen, J., Li, X., Qiao, X. 2010. A comparative study on source apportionment of polycyclic aromatic hydrocarbons in sediments of the Daliao River, China: Positive matrix factorization and factor analysis with non-negative constraints. *Chin. Sci. Bull.* 55; 915–920.

Malamis, S., Andreadakis, A., Mamais, D., Noutsopoulos, C. 2011. Investigation of long-term operation and biomass activity in a membrane bioreactor system. *Water Sci. Technol.* 63; 1906–1912.

Malamis, S., Katsou, E., Di Fabio, S., Frison, N., Cecchi, F., Fatone, F. (2014). Treatment of petrochemical wastewater by employing membrane bioreactors: A case study of effluents discharged to a sensitive water recipient. *Desalination Water Treatment* (ahead of print), 1–10.

Malik, A., Verma, P., Singh, A.K., Singh, K.P. 2011. Distribution of polycyclic aromatic hydrocarbons in water and bed sediments of the Gomti River, India. *Environ. Monit. Assess.* 172; 529–545.

Manoli, E., Samara, C. 1999. Occurrence and mass balance of polycyclic aromatic hydrocarbons in the Thessaloniki sewage treatment plant. *J. Environ. Qual.* 28; 176–187.

Manoli, E., Samara, C. 2008. The removal of polycyclic aromatic hydrocarbons in the wastewater treatment process: Experimental calculations and model predictions. *Environ. Pollut.* 151; 477–485.

Mansuy-Huault, L., Regier, A., Faure, P. 2009. Analyzing hydrocarbons in sewer to help PAH source apportionment in sewage sludge. *Chemosphere*, 75; 995–1002.

Metcalf, L., Eddy, H.P. 2003. *Wastewater engineering treatment and reuse.* New York: McGraw-Hill.

Min, K., Ergas, S. 2006. Volatilization and biodegradation of VOCs in membrane bioreactors (MBR). *Water Air Soil Pollut.* 6; 83–96.

Mozo, I., Stricot, M., Lesage, N., Spérandio, M. 2011. Fate of hazardous aromatic substances in membrane bioreactors. *Water Res.* 45; 4551–4561.

Namkung, E., Rittmann, B.E. 1987. Estimating volatile organic compound emissions from publicly owned treatment works. *J. Water Pollut. Control Fed.* 59; 670–678.

Nikolaou, A.D., Golfinopoulos, S.K., Kostopoulou, M.N., Kolokythas, G.A., Lekkas, T.D. 2002. Determination of volatile organic compounds in surface waters and treated wastewater in Greece. *Water Res.* 36; 2883–2890.

NTUA, National Technical University of Athens, School of Chemical Engineering. Unpublished data from the work of S. Malamis.

Patrolecco, L., Ademollo, N., Capri, S., Pagnotta, R., Polesello, S. 2010. Occurrence of priority hazardous PAHs in water, suspended particulate matter, sediment and common eels (*Anguilla anguilla*) in the urban stretch of the river Tiber (Italy). *Chemosphere* 81; 1386–1392.

Qiao, M., Qi, W., Liu, H., Qu, J. 2014. Occurrence, behavior and removal of typical substituted and parent polycyclic aromatic hydrocarbons in a biological wastewater treatment plant. *Water Res.* 52; 11–19.

Rezouga, F., Hamdi, M., Sperandio, M. 2009. Variability of kinetic parameters due to biomass acclimation: Case of paranitrophenol biodegradation. *Bioresour. Technol.* 100; 5021–5029.

Samara, C., Linelman, J., Kettrup, A. 1995. Determination of selected polynuclear aromatic hydrocarbons in waste water and sludge samples by HPLC with fluorescence detection. *Toxicol. Environ. Chem.* 48; 89–102.

Sánchez-Ávila, J., Bonet, J., Velasco, G., Lacorte, S. 2009. Determination and occurrence of phthalates, alkylphenols, bisphenol A, PBDEs, PCBs and PAHs in an industrial sewage grid discharging to a municipal wastewater treatment plant. *Sci. Total Environ.* 407; 4157–4167.

Stricot, M., Filali, A., Lesage, N., Sperandio, M., Cabassud, C. 2010. Side-stream membrane bioreactors: Influence of stress generated by hydrodynamics on floc structure, supernatant quality and fouling propensity. *Water Res.* 44; 2113–2124.

Stringfellow, W.T., Alvarez-Cohen, L. 1999. Evaluation the relationship between the sorption of PAHs to bacterial biomass and biodegradation. *Water Res.* 33; 2535–2544.

Tian, F., Chen, J., Qiao, X., Wang, Z., Yang, P., Wang, D., Ge, L. 2009. Sources and seasonal variation of atmospheric polycyclic aromatic hydrocarbons in Dalian, China: Factor analysis with non-negative constraints combined with local source fingerprints. *Atmos. Environ.* 43; 2747–2753.

Timoney, K.P., Lee, P. 2011. Polycyclic aromatic hydrocarbons increase in Athabasca River delta sediment: Temporal trends and environmental correlates. *Environ. Sci. Technol.* 45; 4278–4284.

Verlicchi, P., Cattaneo, S., Marciano, F., Masotti, L., Vecchiato, G., Zaffaroni, C. 2011. Efficacy and reliability of upgraded industrial treatment plant at Porto Marghera, near Venice, Italy, in removing nutrients and dangerous micropollutants from petrochemical wastewaters. *Water Environ. Res.* 83; 739–749.

Vogelsang, C., Grung, M., Jantsch, T.G., Tollefsen, K.E., Liltved, H. 2006. Occurrence and removal of selected organic micropollutants at mechanical, chemical and advanced wastewater treatment plants in Norway. *Water Res.* 40; 3559–3570.

Weiss, S., Reemtsma, T. 2008. Membrane bioreactors for municipal wastewater treatment—A viable option to reduce the amount of polar pollutants discharged into surface waters? *Water Res.* 42; 3837–3847.

Wiszniowski, J., Ziembinska, A., Ciesielski, S. 2011. Removal of petroleum pollutants and monitoring of bacterial community structure in a membrane bioreactor. *Chemosphere* 83; 49–56.

Yang, C., Chen, F., Luo, S., Xie, G., Zeng, G., Fan, C. 2010. Effects of surfactants and salt on Henry's constant of n-hexane. *J. Hazard. Mater.* 175; 187–192.

7

PAHs in Wastewater and Removal Efficiency in Conventional Wastewater Treatment Plants

Vincenzo Torretta

Università degli Studi dell'Insubria, Varese, Italy

CONTENTS

7.1　The Behavior of PAHs in Water

This chapter discusses how polycyclic aromatic hydrocarbons (PAHs) can be removed from wastewater by exploiting conventional wastewater treatment plants (WWTPs). The sludge produced during the treatment process is also investigated.

Due to their toxic, mutagenic, and carcinogenic properties, 16 PAHs have been identified as priority pollutants by the U.S. Environmental Protection Agency (USEPA), 7 of which are considered some of the strongest known carcinogenic compounds. Benzo[α]pyrene (BaP) is the most important PAH and, together with dibenzo[a,h]anthracene (dBahA), it is the most toxic PAH (IARC, 1991; Park et al., 2006).

With the exception of naphtalene, PAHs have a high fusion and boiling temperature, and usually a low vapor tension, which is inversely proportional to the number of rings or the molecular mass. The molecular structure determines the stability: a low level of stability indicates a linear structure.

Due to their lipophilicitiy, PAHs can easily cross biological membranes and accumulate inside organisms, causing damage to the genetic material. A measure of molecular lipophilicitiy is represented by the octanol-water (K_{ow}) coefficient, which represents the accumulation capacity in nonpolar phases.

This coefficient is widely used. According to the USEPA (1990), compounds with a $\log K_{ow}$ value higher than 3.5 are potentially dangerous for the environment.

PAHs have a very low solubility in water, though their solubility increases if some organic liquids are present. Solubility decreases with an increase in molecular weight.

The depositing of PAHs in sediments and on particles acts as a reserve, and the PAHs are slowly released back into the water. Thus, soil contamination can cause groundwater contamination. In groundwater it is rare for contamination to be caused by a single PAH or a single hydrocarbon compound; rather, a mix of different pollutants is nearly always found (Nikolau et al., 1984; Schlautman et al., 2004; Zhang et al., 2005; Brandli et al., 2007; Foan et al., 2012; Gunawardena et al., 2012; Jackob et al., 2012; Meynet et al., 2012; Richardson et al., 2012; Scipioni et al., 2012).

7.2 PAHs in Wastewater and Removal

7.2.1 Case Study

If PAHs are not totally mineralized in a treatment system, then some fractions may be released into the surrounding environment in the final effluent, in the sludge or in the atmosphere. The current trend is to evaluate the depurative efficiency of a WWTP on the basis of not only the removal of traditional indicators, but also the abatement of other contaminants (Metcalf and Eddy, 2003).

Because of their hydrophobic behavior, PAHs are adsorbed on organic matter particles and, considering their affinity with particulate, are expected to be significantly removed during primary and secondary sedimentation.

Although there are many studies on PAHs in wastewater and the related removal efficiency in conventional WWTPs, the results are varied (Manoli and Samara, 1999; Blanchard et al., 2001; Blanchard et al., 2004; Katsoyiannis and Samara, 2004; Dai et al., 2006; Katsoyiannis et al., 2006; Hua et al., 2008; Kriipsalu et al., 2008; Xiao et al., 2008; Alhafez et al., 2012; Torretta and Katsoyinnis, 2013). This chapter outlines a case study where the evaluation of PAHs and their removal efficiency in wastewater were carried out at four different WWTPs. All the plants were located in the same district (very industrialized and with a high density of inhabitants) using average values and considering only plants with a prevalent composition of domestic wastewater (60 to 75%) compared to industrial wastewater (however compatible for quality characterization, to be added to domestic wastewaters in conventional treatment).

All four WWTPs, and in particular WWTP3 (100,000 equivalent inhabitants), receive rainwater drained from a very urbanized area, with heavy traffic. Traffic can release an array of harmful pollutants, such as hydrocarbons and heavy metals, into the urban atmosphere. A portion of the traffic-generated pollutants is deposited directly onto the ground surfaces and conducted by the rainfall water into the sewage system.

All four plants are conventional biological treatment plants with mechanical pretreatments (gridding, sand/oil removal, and primary sedimentation), followed by biological treatment (activated sludge) and final sedimentation and disinfection. The sludge line is complete for all the plants, with anaerobic digestion, thickening, conditioning, and final dewatering.

Characteristics and processes of the four WWTPs are given in Tables 7.1 and 7.2.

TABLE 7.1

Characteristics of the Four WWTPs—Average Values

Parameter	Unit	WWTP1	WWTP2	WWTP3	WWTP4
Equivalent inhabitants		80,000	400,000	100,000	25,000
Inflow	$(m^3 \ day^{-1})$	22,000	100,000	27,000	5500
% urban	%	75	70	60	65
% industrial	%	25	30	40	35

TABLE 7.2

Description of the Four WWTPs

Line	Process	WWTP1	WWTP2	WWTP3	WWTP4
Water line	Mechanical pretreatment (gridding and sand/oil removal)	X	X	X	X
	Primary sedimentation		X	X	
	Biological treatment (activated sludge)	X	X	X	X
	Final sedimentation	X	X	X	X
	Disinfection	X	X	X	X
Sludge line	Prethickening	X	X	X	X
	Anaerobic digestion	X	X	X	X
	Post-thickening		X		
	Conditioning	X	X	X	X
	Dewatering	X	X	X	X

In order to estimate the removal efficiency of PAHs in conventional WWTPs, this case study investigates the presence of PAHs in wastewater over a period of 1 year of observations and samplings. The concentrations measured were below the legal limits; however, in a few cases statistical analysis revealed correlations between various chemical-physical indicators, such as $\log K_{ow}$ and $\log K_H$, and PAH removal yield, verified at various stages of the different treatment plants. These correlations help to assess the level of danger associated with the water effluents of a treatment plant in relation to the values of some of the statistical indicators considered here.

7.2.2 Samplings

Throughout the 1-year survey period, two different samples were collected on each day at three different locations in the plant (considering the hydraulic retention time existing between the sampling locations): position A just after the gridding device, i.e., where the process begins; position B after the primary sedimentation or mechanical treatment; position C after the disinfection device, i.e., the end of the process (water line).

The samples were homogenized, lyophilized, and purified before gas chromatography–mass spectrometry (GC-MS) analyses.

Total PAHs (adsorbed and dissolved) were extracted by the liquid-liquid method and analyzed by high-performance liquid chromatography (HPLC) and fluorescence revelations. The samples were purified using silica chromatographic columns and analyzed by gas chromatography coupled with mass spectrometry. All solvents and reagents were of chromatographic grade.

At the beginning, we considered different types of PAHs, but the low concentration levels focused our attention on a smaller group of compounds. The initial group was also considered in light of the literature, and consisted of the following:

Naphthalene
Acenaphthylene
Acenaphthene
Fluorene
Phenanthrene
Anthracene
Fluoranthene
Pyrene
Benz[α]anthracene
Chrysene
Benzo[b]fluoranthene
Benzo[k]fluoranthene

Benzo[α]pyrene

Dibenzo[α,h]anthracene

Benzo[g,h,i]perylene

Indeno[1,2,3-c,d]pyrene

The restricted group of target compounds had significant concentration values. It included anthracene, fluoranthene, pyrene, chrysene, and benz[α]anthracene. In addition, naphthalene, phenanthrene, benzo[b]fluoranthene, benzo[α] pyrene, benzo[g,h,i]perylene, benzo[k]fluoranthene, and indeno[1,2,3-c,d] pyrene were target compounds, but were always below the detection limits.

For each PAH the average concentration value was calculated, measured in the four WWTPs at the corresponding treatment stages. The lowest values were rejected.

Thus for each PAH, three concentrations were identified by evaluating the corresponding removal yields: the average concentration at the entrance of the WWTP, the average after the primary sedimentation (mechanical treatment), and the average after the secondary sedimentation (or disinfection).

7.2.3 Theoretical Removal Rate by FATE Model

To evaluate the PAH removal in a traditional WWTP, the FATE (Fate and Treatability Estimator) model is frequently applied, which was developed by the Technology Industrial Division of the USEPA in 1990.

The model is made up of two submodels: one to evaluate the removal of the organic fraction and the other for the inorganic fraction (Byrns, 2001; Manoli and Samara, 2008; Torretta, 2012).

The organic submodel takes into consideration the primary sedimentation, the aeration, and the secondary sedimentation. It accounts for the following removal processes: primary phase adsorption and secondary phase adsorption, volatilization, and biodegradation. The outflow concentration from the primary sedimentation is hypothesized as being equal to the incoming concentration in the aeration tank and secondary sedimentation. The outflow pollutant concentrations from primary and secondary sedimentation are shown in Equations 7.1 and 7.2:

EQUATIONS 7.1 AND 7.2

Outflow pollutant concentrations from primary (1) and secondary (2) sedimentation. (From USEPA, Cercla Site Discharges to POTWS: Treatability Manual, EPA 540/2-90/007, Washington, DC, 1990, pp. 522–540.)

$$S_0 = QS_{in} / Q + Q_p X_p \left(4.1 \cdot 10^{-5}\right) K_{ow}^{0.35} \tag{7.1}$$

$$S = \left(QS_0\right) / \left(Q + GH/RT + Q_w X_v\right)\left(3.06 \cdot 10^6 K_{ow}^{0.67}\right) + K_1 X_a V \tag{7.2}$$

$Q = Q_0$ = incoming flow rate = outflow flow rate from secondary sedimentation (Q_c water + Q_w sludge), S_0 = organic pollutant concentration outflow from primary sedimentation, S_{in} = incoming organic pollutant concentration at WWTP, Q_p = sludge flow rate extracted from primary sedimentation, Q_w = sludge flow rate extracted from secondary sedimentation, X_p = dry substance concentration in primary sedimentation (%), X_v = pollutant concentration in secondary sludge, X_a = active cell concentration in the biological reactor (assumed as 0.64 of the SS in the mixed liquor), G = air flow rate in aeration compartment, H = Henry's law constant, $R = 8.206 \times 10^5$ (m^3 atm K^{-1} mol^{-1}), T = aeration compartment temperature, K_1 = first-order biodegradation coefficient, and K_{ow} = water/octanol coefficient.

As with the measured data, once the outflow concentrations had been determined, the removal percentage values were calculated according to the model.

7.2.4 Comparison between Observed and Theoretical Data

The observed values, obtained from the analysis, were compared with the values obtained from the FATE model. Figure 7.1 shows the concentration averages in the wastewater inflow (stage A), after the primary sedimentation (stage B), and after the secondary sedimentation (stage C), along with the corresponding removal yields.

We found that not all 16 PAHs that the technical literature considers as the main pollutants in the group are present in a significant quantity in the

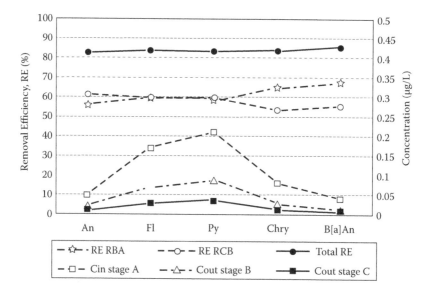

FIGURE 7.1
Removal efficiencies and concentrations for anthracene (An), fluoranthene (Fl), pyrene (Py), chrysene (Chry), and benz[α]anthracene (B[α]An). Values shown are rounded averages for the four WWTPs.

TABLE 7.3

Comparison between Observed and Calculated Removal Percentages

Chemical	Removal Efficiency R_1, (%)		Total Removal Efficiency R_{tot}, (%)	
	Observed	Calculated	Observed	Calculated
Anthracene	56.0	53.0	83.0	79.8
Fluoranthene	59.4	58.2	83.5	81.6
Pyrene	58.6	57.6	83.3	80.6
Chrysene	65.0	62.5	83.8	82.4
Benzo[α] anthracene	67.5	67.0	85.6	84.6

Note: R_1 = removal efficiency after mechanical stage.

wastewater. In fact, only five PAHs are on average present in the influent of all four WWTPs, with values between 0.040 µg L^{-1} (benz[α]anthracene) and 0.210 µg L^{-1} (pyrene).

Although we are considering low concentration values, the percentage of removal related to the mechanical treatments is nevertheless significant, with yields in the range of 56% (anthracene) and 67.5% (benz[α]anthracene), see Table 7.3.

Considering the secondary treatment, in all the cases characterized by the activated sludge process, the removal yields range between 53.5% (chrysene) and 61.3% (anthracene), showing an opposite trend compared with the mechanical treatment. For example, the primary treatment is less efficient regarding anthracene removal, whereas the secondary treatment is the most efficient in terms of anthracene abatement. Obviously, the results need to be considered with due care, given the quite low values of the concentrations.

Using the average values from all four WWTPs, a good correlation was found between the average concentration values and concentrations analyzed using FATE. The results show how after primary sedimentation, there were no differences between the experimental (real) values and those calculated by FATE. The maximum difference in percentage between the observed removal and calculated removal was approximately 2 to 3%. Also in the final effluent, for some PAHs there was good agreement between the theoretical and experimental results; in other cases, FATE underestimated the final results.

7.2.5 Statistical Correlations

In order to examine a possible statistical correlation with various indicators of the physical-chemical compound properties, such as $logK_{ow}$ and $logK_H$, the second step involved using the removal yields calculated after the primary treatment and after the biological stage.

In the mechanical treatment, compounds with a $\log K_{ow}$ value higher than 5 have a removal yield of between 60 and 66%, whereas compounds with a $\log K_{ow}$ value higher than 3.6 and lower than 5 have a less than 60% removal efficiency (Table 7.4). This is important because compounds with a higher than 3.5 $\log K_{ow}$ value are potentially dangerous for the environment and must be removed during mechanical treatments. It confirms the risks associated with not considering the mechanical treatment when, for example, a higher value of organic load is required in the biological treatment in order to improve the denitrification. This critical aspect needs to be taken into account when evaluating the possible consequences to PAH removal of neglecting the mechanical treatment.

Within the activated sludge, the injection of diffused air can induce a transfer to the atmosphere. As volatilization is strongly connected to Henry's law constant, a potential correlation can be found between the biological removal yield and $\log K_H$.

In this case, too, the statistical evaluations revealed a good correlation (Table 7.5), in line with the literature.

TABLE 7.4

Correlation between Removal Efficiency after Primary Treatment Stage and $\log K_{ow}$

Chemical	$\log K_{ow}$ (–)	R_1 (%)
Anthracene	4.45	56.0
Fluoranthene	4.90	59.4
Pyrene	4.88	58.6
Chrysene	5.61	65.0
Benzo[α]anthracene	5.61	67.5

Note: Linear regression equation: $R = 9.191 \log K_{ow} + 14.517$; $R^2 = 0.9542$.

TABLE 7.5

Correlation between Removal Efficiency after Secondary Treatment Stage (R_2) and $\log K_H$

Chemical	$\log K_H$ (–)	R_2 (%)
Anthracene	–4.47	61.4
Fluoranthene	–5.18	59.4
Pyrene	–4.92	59.8
Chrysene	–5.97	53.6
Benzo[α]anthracene	–5.82	55.6

Note: Linear regression equation: $R_2 = 5.022 \log K_H + 84.422$; $R^2 = 0.9441$.

7.3 PAH in Sludge: Case Study

Sewage sludge results from the sedimentation of suspended solids during wastewater treatment. Within the European Union alone, more than 10 million tons of sewage sludge is produced every year. Due to the amount produced and the accumulated pollution, sewage sludge has a critical impact on the environment. The main disposal routes of sewage sludge include incineration, sanitary landfill, or use in agricultural land as a soil amendment. In sewage sludge (for use as soil amendment), the presence of PAHs has been regulated and the sum of 16 USEPA PAHs should not exceed the EU-set target concentration of 6 mg kg^{-1} (dry matter).

Most harmful hydrophobic organic contaminants in wastewaters tend to sorb on suspended solids, thus becoming sewage sludge, resulting in over 1000 chemicals from a diverse range of classes of compounds being identified in sewage sludge. Contaminant concentrations vary from the pg kg^{-1} to g kg^{-1} range, depending mainly on the quality and type of incoming wastewater (domestic, municipal, industrial, etc.). The most common hydrophobic organic contaminants are PAHs, polychlorinated biphenyls (PCBs), phthalates, polychlorinated dibenzo-*p*-dioxins and furans (PCDD/Fs), organochlorine insecticides, chlorobenzenes, amines, nitrosamines, and phenols.

The case study was carried out in one of the four WWTPs described above, with the highest capacity: 400,000 Equivalent Inhabitants (E.I.).

7.3.1 Sampling and Analysis

Again, the observation period was about 1 year. The samples were collected every month from the primary sludge (PS), secondary sludge (SS), after the prethickening (PrT), and after the dewatering stage (final sludge, FS), and were bulked together to give an average monthly sample.

In brief, the sludge samples were ultrasonically extracted with an aliquot of hexane-dichloromethane, further purified using silica chromatographic columns, and analyzed by means of gas chromatography coupled with mass spectrometry. All solvents and reagents were of chromatographic grade. The target compounds were:

Naphthalene

Phenanthrene

Anthracene

Fluoranthene

Pyrene

Benz[α]anthracene

Chrysene

Benzo[b]fluoranthene

Benzo[α]pyrene (BaP)

Benzo[g,h,i]perylene

In addition, benzo[k]fluoranthene and indeno[1,2,3-c,d]pyrene were investigated, but were always below the limit of detection.

7.3.2 PAHs in the Sludge Line

Several PAHs were never detected; therefore, we present only the concentrations of PAHs that were consistently present. PAHs tend to sorb on particles, thereby depositing themselves in the PS. In this study, average individual PAHs varied between 26.7 ng g^{-1} for anthracene and 480 ng g^{-1} for pyrene. The average sum of PAHs was 2500 ng g^{-1}. BaP was on average 56.6 ng g^{-1}, accounting for 2% of the total PAHs. By analyzing PAHs throughout the whole year, it was shown that the PAH concentrations were quite constant, with an increase in the summer months, which is contrary to the known seasonality of PAHs—thus their tendency to exhibit higher concentrations in the winter. This suggests that the PAHs were mainly due to traffic sources and industrial emissions, but it is unlikely that the main sources were related to domestic or industrial heating. ΣPAHs varied between 2325 ng g^{-1} (January) and 2890 ng g^{-1} (July).

Similar to PS, the lowest average concentration in the SS was for anthracene (23.4 ng g^{-1}), and the highest for pyrene (429 ng g^{-1}). The highest concentration was again observed for the July samples, confirming the influence of seasonality. In general, the concentrations were constant and the relative standard deviation between the various monthly samples was 11%. This is probably due to the fact that monthly samples are bulks of several subsamples. Compared to PS, concentrations in the SS were higher in 10 out of 12 cases. This suggests that the primary treatment did not remove the PAHs quantitatively.

During PrT, a large increase in the concentration of suspended solids takes place, but the concentrations of PAHs were not found to be affected. Anthracene was again the least abundant (22.9 ng g^{-1}), and pyrene had the highest average concentrations (472 ng g^{-1}).

ΣPAHs ranged between 2015 ng g^{-1} (June) and 2645 ng g^{-1} (August), which is the same order of magnitude as in PS and SS. This shows the resistance of PAHs in this specific treatment process (PrT), as all PAHs remained intact. The fact that the concentration of PAHs is at the same order of magnitude throughout the whole sludge line suggests that the treatment steps satisfactorily remove the water content, but do not have any degradation effect on PAHs. It also appears that the volatile suspended solids are not a large fraction of the total suspended solids. Correlation analysis of ΣPAHs between the first three steps was particularly weak, probably due to the mixing of PS and SS. The SS contained PAHs in the same order of magnitude; however, this weak correlation is probably due to wet deposition during rain, which can significantly change the PAH profile of the various types of sludge.

During the anaerobic digestion and post-thickening steps that lead to the FS of this specific WWTP, the water content is further removed; the concentrations of PAHs, however, remain basically the same. The levels of concentrations are constant, with small differences compared to the earlier stages. Differences in the profiles are attributed to the fact that the FS is a mixed-composite sample from more than 21 days; thus, minor concentration differences are expected. ΣPAHs in the FS were again in the order of 3000 ng g^{-1} dry weight, which is about half of the maximum permissible limit set by the EU for the safe use of sewage sludge in agriculture as soil amendment. BaP never exceeded 100 ng g^{-1}, which was at the lowest concentration levels found in the literature.

7.3.3 Mass Balance of PAHs through the Sludge Line

The mass balance of total PAHs was calculated in the sludge line, using the average flow rates for each step, the concentrations of suspended solids, and the PAH concentrations. The average daily inputs of PAHs through the primary and the secondary sludge match well with the average daily content of PAHs in the prethickened sludge. In the cases of anthracene, benzo[b]fluoranthene, benzo[α]pyrene, and benzo[g,h,i]perylene, the daily PAH content in PS and SS was slightly higher than in the PrT sludge. The mass balance closures were considered very satisfactory, and the small differences (always below 10%) can be attributed to the mismatching of samples. The mass balances, however, were not satisfactory between the FS and the PrT sludge. In this case, the calculated daily content of PAHs in the FS was far higher than in the PrT sludge (average 22.8%), while in the case of benz[α]anthracene, the FS content was 40% higher. The FS is a representative composite of the monthly sample, so here the effect of sample mismatching may be larger. Another explanation is possibly inaccurate data concerning the daily production of sewage sludge, provided by the wastewater treatment plant authorities. The total annual PAH output was around 30 kg of net PAHs in the sewage sludge alone.

7.3.4 Sewage Sludge as a Predictor for Raw Wastewater Concentrations

Based on the sewage sludge concentrations, we attempted to calculate the raw wastewater concentrations. One way to do this is by using the sludge concentrator factor (SCF), calculated by the concentrations of PAHs in raw wastewater and sludge, as shown in Equation 7.3:

EQUATION 7.3
Sludge concentrator factor calculation.

$$SCF = \frac{C_{Sl}}{C_{Ww}} \tag{7.3}$$

C_{Sl} is the concentration in FS (ng kg^{-1}), and C_{Ww} is the concentration in raw wastewater (ng L^{-1}).

For compounds with $\log K_{ow}$ higher than 5, SCF is approximately 1300 (L kg^{-1}). For chemicals with a $\log K_{ow}$ around 4, SCF is on the order of 800, and for compounds with a $\log K_{ow}$ of 3, SCF is approximately 100. Based on the average FS concentrations, the average annual wastewater concentrations in the studied WWTP should range between 21.4 ng L^{-1} for anthracene and 668 ng L^{-1} for naphthalene. The latter was not among the most abundant in FS, but its lower SCF suggests that a significant part of naphthalene remained in the wastewater stream. A comparison of the findings of this approach, with the annual average wastewater results provided by the authorities of the WWTP in question, highlights that the SCF approach provides results that are in the same order of magnitude; given the sampling mismatch, these estimated results can be considered satisfactory.

7.4 Correlation between PAHs and Industrial Flows

As a last step, a specific PAH measurement was taken at the end of the sludge treatment line in order to evaluate any correlation between PAH concentration and the percentage of industrial wastewater inflow. The monthly average values were considered regarding the four WWTPs at the end of the treatment (after dewatering). The average values are reported in Figure 7.2.

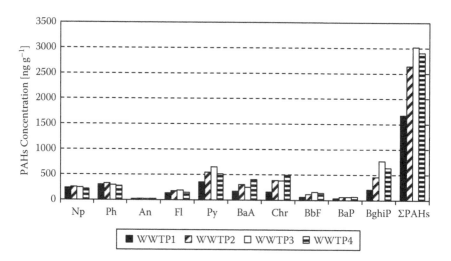

FIGURE 7.2
Concentrations of PAHs in the sludge samples, final stage (after dewatering). PAHs: naphthalene (Np), phenanthrene (Ph), anthracene (An), fluoranthene (Fl), pyrene (Py), benz[α]anthracene (BaA), chrysene (Chr), benzo[b]fluoranthene (BbF), benzo[α]pyrene (BaP), benzo[g,h,i] perylene (BghiP).

TABLE 7.6

Concentrations of PAHs and Industrial
Wastewater Incidence

Industrial Wastewater Proportion (%)	ΣPAHs Concentration (ng g^{-1})
25	1675
30	2645
35	2904
40	3020

Note: Linear regression equation: $c_{PAH} = 85.880$ industrial wastewater incidence (%) – 230.10; $R^2 = 0.82$.

In dewatered sludge the presence of pyrene was on average higher than that of the other PAHs. In fact, pyrene represents 20% of the total amount of PAHs in the dehydrated sludge, whereas anthracene has the least significant presence (<1%).

Given that we were investigating PAHs in sludge, we focused on the correspondence of PAHs with the percentage of industrial components in the wastewater inflow. In fact, the proportion of industrial wastewater flowing into the WWTPs significantly correlated with the total content of PAHs in the sewage sludge (Table 7.6). This suggests that the proportion of industrial wastewater plays an important role in the total concentration of PAHs in the sludge from different WWTPs.

The correlations that we found in the four WWTPs (both in water lines and in the sludge lines) suggest that the chemical and physical indicators outlined in this chapter give an accurate picture of the presence of PAHs in WWTPs and in the related effluents. We thus believe that our results can be exploited to develop accurate controls of PAHs in the sewage treated at WWTPs, thereby ensuring higher removal rates. The environment would thus be better protected, and the risk to human health would be reduced considerably.

References

Alhafez L., Muntean N., Muntean E., Mihaiescu T., Mihaiescu R., Ristoiu D. 2012. Polycyclic aromatic hydrocarbons in wastewater sewerage system from the Cluj-Napoca area. *Environ. Eng. Manage. J.*, 11(1), 5–12.

Blanchard M., Teil M.J., Ollivon D., Garban B., Chesterikoff C., Chevreuil M. 2001. Origin and distribution of polyaromatic hydrocarbons and polychlorinated biphenyls in urban effluents to wastewater treatment plants of the Paris area (France). *Water Res.*, 35, 3679–3687.

Blanchard M., Teil M.J., Ollivon D., Legenti L., Chevreuil M. 2004. Polycyclic aromatic hydrocarbons and polychlorobiphenyls in wastewaters and sewage sludges from the Paris area (France). *Environ. Res.*, 95, 184–197.

Brandli R., Bucheli T.D., Kupper T., Mayer J., Stadelmann F.X., Taradellas J. 2007. Fate of PCBs, PAHs and their source characteristic ratios during composting and digestion of source-separated organic waste in full-scale plants. *Environ. Pollut.*, 148(2), 520–528.

Byrns G. 2001. The fate of xenobiotic organic compounds in wastewater treatment plants. *Water Res.*, 35(10), 2523–2533.

Dai J., Xu M., Chen J., Yang X., Ke Z. 2006. PCDD/F, PAH and heavy metals in the sewage sludge from six wastewater treatment plants in Beijing, China, *Chemosphere*, 66, 353–361.

Foan L., Domerq M., Bermejo R., Santamaria J.M., Simon V. 2012. Polycyclic aromatic hydrocarbons (PAHs) in remote bulk and throughfall deposition: Seasonal and spatial trends. *Environ. Eng. Manage. J.*, 11(5), 1101–1111.

Gunawardena J., Egodawatta P., Ayoko G.A., Goonetilleke A. 2012. Role of traffic in atmospheric accumulation of heavy metals and polycyclic aromatic hydrocarbons. *Atmos. Environ.*, 54, 502–510.

Hua L., Wu W.X., Liu Y.X., Tientchen C.M., Chen Y.X. 2008. Heavy metals and PAHs in sewage sludge from twelve wastewater treatment plants in Zhejiang Province. *Biomed. Environ. Sci.*, 21, 345–352.

IARC (International Agency for Research on Cancer). 1991. *Monographs on the evaluation of carcinogenic risks to humans*. Lyon, France, IARC, pp. 43–53.

Jakob L., Hartnik T., Henriksen T., Elmquist M., Brändli R.C., Hale S.E., Cornelissen G. 2012. PAH-sequestration capacity of granular and powder activated carbon amendments in soil, and their effects on earthworms and plants. *Chemosphere*, 88, 699–705.

Katsoyiannis A., Samara C. 2004. Persistent organic pollutants (POPs) in the sewage treatment plant of Thessaloniki, northern Greece: Occurrence and removal. *Water Res.*, 38, 2685–2698.

Katsoyiannis A., Zouboulis A., Samara C. 2006. Persistent organic pollutants (POPs) in the conventional activated sludge treatment process: Model predictions against experimental values. *Chemosphere*, 65, 1634–1641.

Kriipsalu M., Marques M., Hogland W., Nammari D.R. 2008. Fate of polycyclic aromatic hydrocarbons during composting of oily sludge. *Environ. Technol.*, 29, 43–53.

Manoli E., Samara C. 1999. Occurrence and mass balance of polycyclic aromatic hydrocarbons in the Thesssalonoki sewage treatment plant. *J. Environ. Qual.*, 28, 176–187.

Manoli E., Samara C. 2008. The removal of polycyclic aromatic hydrocarbons in the wastewater treatment process: Experimental calculations and model predictions. *Environ. Pollut.*, 151, 477–485.

Metcalf L., Eddy H.P. 2003. *Wastewater engineering—Treatment and reuse*, 4th ed. McGraw-Hill, New York.

Meynet P., Hale S.H., Davenport R.J., Cornelissen G., Breedveld G.D., Werner D. 2012. Effect of activated carbon amendment on bacterial community structure and functions in a PAH impacted urban soil. *Environ. Sci. Technol.*, 46, 5057–5066.

Nikolaou K., Masclet P., Mouvier G. 1984. Sources and chemical reactivity of polynuclear aromatic hydrocarbons in the atmosphere: A critical review. *Sci. Total Environ.*, 32, 103–132.

Park S.-Y., Lee K.-H., Kang D., Lee K.-H., Ha E.-H., Hong Y.-C. 2006. Effect of genetic polymorphisms of MnSOD and MPO on the relationship between PAH exposure and oxidative DNA damage. *Mutat. Res.*, 593(1–2), 108–115.

Richardson S.D., Jones M.D., Singleton D.R., Aitken M.D. 2012. Long-term simulation of in situ biostimulation of polycyclic aromatic hydrocarbon-contaminated soil. *Biodegradation*, 23, 621–633.

Schlautman M.A., Yim S., Carraway E.R., Lee J.H., Herbert B.E. 2004. Testing a surface tension-based model to predict the salting out of polycyclic aromatic hydrocarbons in model environmental solutions. *Water Res.*, 38, 3331–3339.

Scipioni C., Villanueva F., Pozo K., Mabilia R. 2012. Preliminary characterization of polycyclic aromatic hydrocarbons, nitrated polycyclic aromatic hydrocarbons and polychlorinated dibenzo-p-dioxins and furans in atmospheric PM10 of an urban and a remote area of Chile. *Environ. Technol.*, 33, 809–820.

Torretta V. 2012. PAH in wastewater: Removal efficiency in a conventional wastewater treatment plant and comparison with model predictions. *Environ. Technol.*, 33, 851–855.

Torretta V., Katsoyiannis A. 2013. Occurrence of polycyclic aromatic hydrocarbons in sludges from different stages of a wastewater treatment plant in Italy. *Environ. Technol.*, 34(7), 937–943.

USEPA. 1990. *Cercla site discharges to POTWS: Treatability manual*. EPA 540/2-90-007. Washington, DC.

Xiao R., Du X., He H., Zhang Y., Yi Z., Li F. 2008. Vertical distribution of polycyclic aromatic hydrocarbons (PAHs) in Hunpu wastewater-irrigated area in northeast China under different land use patterns. *Environ. Monit. Assess.*, 142, 23–34.

Zhang X.L., Tao S., Liu W.X., Yang Y., Zuo Q., Liu S.Z. 2005. Source diagnostics of polycyclic aromatic hydrocarbons based on species ratios. A multimedia approach. *Environ. Sci. Technol.*, 39, 9109–9114.

8

PAHs in Wastewater during
Dry and Wet Weather

Kenya L. Goodson,[1] Robert Pitt,[2] and Shirley Clark[3]

[1]Nspiregreen, Washington, DC

[2]Department of Civil, Construction, and Environmental Engineering,
University of Alabama, Tuscaloosa, Alabama

[3]Penn State Harrisburg, Middletown, Pennsylvania

CONTENTS

8.1 Introduction

8.1.1 PAHs: Definition and Chemical Properties

Polycyclic aromatic hydrocarbons (PAHs) are ringed hydrocarbons that are created during the incomplete combustion of petroleum fuels. They also naturally occur as major components of petroleum. They are chemicals of concern because of their mutagenic and carcinogenic properties. PAHs are ubiquitous in the modern environment because of widespread petroleum use. Sixteen PAHs are listed as U.S. Environmental Protection Agency (USEPA) priority pollutants, as shown in Table 8.1.

TABLE 8.1

EPA List of 16 Priority PAH Pollutants

Naphthalene	Benz[a]anthracene
Acenaphthylene	Chrysene
Acenaphthene	Benzo[b]fluoranthene
Fluorene	Benzo[k]fluoranthene
Phenanthrene	Benzo[a]pyrene
Anthracene	Dibenz[a,h]anthracene
Fluoranthene	Benzo[g,h,i]perylene
Pyrene	Indeno[1,2,3-c,d]pyrene

Source: USEPA, Polycyclic Aromatic Hydrocarbons, EPA Fact Sheet, Washington, DC, January 2008.

Although the health effects of individual PAHs vary, these PAHs are considered a single toxicological group (ATSDR 2011). These 16 PAHs are specifically listed because more information is available on these compounds than for other PAHs; they are suspected to be more harmful than some of the other PAHs, and they exhibit harmful effects that are representative of the PAHs; there is a greater chance that exposure is more likely to these PAHs than to the others; and of all the PAHs analyzed, these were the PAHs identified at the highest concentrations at National Priority List hazardous waste sites (ATSDR 2011).

In urban areas, PAHs are emitted from vehicle exhaust, domestic heaters, fossil fuel power plants, and industries as a result of hydrocarbon fuel combustion or during the processing of raw materials (Manoli and Samara 1999). Some are released directly into the atmosphere, which then affects ground surfaces or water bodies through atmospheric deposition. The atmospheric deposition of PAHs on urban surfaces can then contaminate storm water during rains. Storm water is a major contributor of PAHs to urban receiving waters. Storm water can also affect wet weather sanitary sewage flows through infiltration and inflow (I&I) or combined sewers.

PAHs are generally insoluble and very lipophilic. They are differentiated by the number of carbon rings and the placement of hydrocarbons along those rings. These structures, specifically the connectedness of the rings, affect many of their chemical properties, such as solubility (Streitweiser et al. 1998). The solubility of some PAHs, such as anthracene (the least soluble PAH shown in Table 8.2), increases with increases of temperature (Andersson et al. 2005). Theoretically, PAHs tend to adsorb onto particulate matter, such as total suspended solids, as predicted by the log10 Kow values (Mackay 2001). This increases their likelihood for removal as the particulates are removed. Based solely on chemical and physical properties of PAHs, sedimentation processes should therefore significantly reduce PAHs in wastewater. Table 8.2 lists some of the physical and chemical properties that affect the fate and transport of PAHs in storm water and sanitary sewer systems.

TABLE 8.2

Chemical Properties of Selected PAHs

Chemical Name (PAH)	Molecular Weight (g/mol)	Solubility (mg/L)	Log Octanol-Water Coefficient (log10 Kow)
Naphthalene	128.2	31.5	3.37
Acenaphthylene	152.2	3.80	3.89
Acenaphthene	154.2	16.1	4.02
Fluorene	166.2	1.90	4.12
Anthracene	178.2	0.045	4.53
Phenanthrene	178.2	1.12	4.48

Source: Goodson et al., *Proceedings of the Water Environment Federation* 2012(7): 7224–37, 2012.

8.1.2 Treatability of PAHs in Conventional Wastewater Treatment Plants

Wastewater treatment plants (WWTPs) are designed to remove conventional pollutants, such as biodegradable organic compounds measured as 5-day biochemical oxygen demand (BOD_5), fecal coliforms, and total suspended solids (TSS). Many plants also have the ability to reduce wastewater nutrient concentrations through the incorporation of the nutrients into the microbial biomass. For those compounds, such as PAHs, that are less readily biodegradable, treatment may require advanced unit processes, such as air stripping or adsorption, which are not commonly found at conventional WWTPs (Tchobanoglous et al. 2003).

Studies have investigated PAH treatment at wastewater treatment facilities, and have included plants that treat separate sanitary sewage flows and combined sewer flows. Combined sewer flows are relevant to this project because those plants also receive highly variable flows, similar to those received by systems influenced by I&I (Phillips and Chalmers, 2009). Blanchard et al. (2001) evaluated samples collected from five combined sewer WWTPs near Paris, France, as a function of rainfall. The results showed that PAH concentrations in the sewage increased during rain events, compared to dry events (see Figure 8.1). These results, although they are based on a very small sample set, indicate that PAHs may be entering the system through the storm water component of the flow.

Pham and Proulx (1997) collected samples from the Montreal (Canada) Urban Community (MUC) wastewater treatment system. The MUC WWTP treats combined domestic, industrial, and storm water wastewaters. Pham's study correlated the treatment effectiveness with the physical and chemical properties of the compounds studied. Naphthalene had the lowest removal rate and also has the lowest adsorption capacity and the highest solubility. Removal rates for PAHs in the treatment plant correlated well with the log Kow of the compounds and with removal rates seen in other studies, especially for the medium-weight PAHs. Table 8.3 highlights the influent and effluent concentrations for the compounds studied.

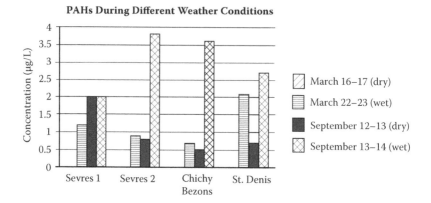

FIGURE 8.1
PAH removal at wastewater treatment facilities in Paris. (Reprinted with permission from Blanchard et al., *Water Research*, 35(15), 3679–87, 2001.)

TABLE 8.3

PAH Removals at a Wastewater Treatment Facility at Montreal

| | N = 10 | | N = 6 | | |
PAH	Avg. Influent, μg/L	Std. Influent, μg/L	Avg. Effluent, μg/L	Std. Effluent, μg/L	Average Removal Rates, %
Naphthalene	0.147	0.084	0.088	0.049	40
Acenaphthylene	0.021	0.051	0.002	0.005	90
Acenaphthene	0.016	0.011	0.005	0.003	67
Fluorene	0.037	0.025	0.015	0.008	59
Phenanthrene	0.333	0.228	0.109	0.055	67
Anthracene	0.028	0.034	0.012	0.007	58
Fluoranthene	0.150	0.193	0.020	0.007	86
Pyrene	0.138	0.157	0.023	0.007	83
Chrysene	0.080	0.122	0.005	0.002	93

Source: Pham and Proulx, *Water Research* 31(8): 1887–96, 1997.

Manoli and Samara (1999) collected composite samples over a 5-day period during both cold and warm weather from the combined sewage WWTP at Thessaloniki, Greece. It is a conventional activated sludge treatment facility that includes the addition of a flocculent and chlorine dioxide for disinfection. The plant receives a dry weather flow of approximately 40,000 m3/day, consisting mainly of residential wastewater. The treatment unit processes include (1) pretreatment with aerated sands and grease removal units, (2) a primary sedimentation tank with a detention time of 3 hours, (3) an aeration tank with surface aerators with a detention time of 3 hours, and (4) a secondary sedimentation tank with a detention time of 6 hours. PAHs analysis was performed for naphthalene, acenaphthene, fluorene, phenanthrene, pyrene, benzo[a]anthracene, and chrysene.

Naphthalene, as reported previously, had the lowest removal rate (32%), followed by fluorene (67%), acenaphthene (84%), pyrene (87%), and phenanthrene (88%). Naphthalene also showed a slight increase in the primary effluent before its reduction in the secondary treatment process, and minor reduction in the final treatment process. Benzo[a]anthracene (91%) and chrysene (91%) had the highest removal rates. Most of these two compounds were removed during primary sedimentation treatment. These compounds are high in molecular weight, implying strong sorption to particulates that are captured during the primary treatment stage. In general, the lower molecular weight PAHs were more likely to be removed in the secondary treatment part of the facility, at least in part due to biodegradation; both naphthalene and acenaphthene saw the bulk of their removal in the secondary treatment. However, phenanthrene had high removals during the secondary treatment process, indicating either that the primary process did not have sufficient residence time for sorbed particle removal or that phenanthrene underwent more biodegradation than has been seen in other studies.

8.2 Impact of Flow Rate on PAH Removal: Case Study in Tuscaloosa, Alabama

Goodson et al. (2012) investigated the impact of wet weather flows, measured as increased flow rates, on PAH removals at a Tuscaloosa, Alabama, wastewater treatment facility. Wastewater samples were collected at four sampling locations within the WWTP during both dry and wet weather to identify the impacts of flow rate changes on removal and determine which unit processes were most effective under this range of flow conditions. Samples were analyzed from the inlet of the treatment plant, after primary treatment, after secondary treatment, and at the effluent to the plant. Theoretical removals of 10 PAHs (a subset of the 16 listed above) based on their physical and chemical properties were compared to observed removals during these different unit treatment processes.

8.2.1 Site Description

The Hilliard N. Fletcher WWTP is located in Tuscaloosa, Alabama; the city has a population of approximately 90,500 according to the 2010 U.S. Census, and a total area of 173 km^2 (27 km^2 of this is Lake Tuscaloosa and the Black Warrior River). The population density is about 620 people/km^2, excluding the water area. Lake Tuscaloosa is the source of Tuscaloosa's drinking water.

The Tuscaloosa wastewater treatment system discharges its effluent into the Black Warrior River and Crib Mills Creek. The winter seasons are generally mild, with temperatures between –7 and 10°C, and the average monthly rainfall depths are about 13 cm. Spring seasons have temperatures between 10 and 27°C, and have similar rainfall depths of 13 cm. Summer temperatures range from 15 to 32°C and can reach 38°C; average monthly rainfall depths are approximately 10–13 cm.

The Hilliard N. Fletcher Wastewater Treatment System is a conventional municipal wastewater treatment facility that uses activated sludge biological treatment. This system includes approximately 885 km of publicly maintained separate sanitary sewers with another 80 km of privately owned sanitary sewers. Over 60 pump stations are used to transport the wastewater to the WWTP. A major expansion (US$33 million) was started in 1995, increasing the capacity of the treatment plant to 91,000 m^3/day. The treatment facility was recently expanded to a capacity of 151,000 m^3/day in 2013. According to the National Pollutant Discharge Elimination System (NPDES) permit, this treatment system services a population of approximately 110,000 (including some of the surrounding communities). The service area is estimated to be about 192 km^2. It is a separate sanitary treatment system and is not designed for storm water treatment or combined flows. There are also industrial discharges entering the Tuscaloosa treatment facility.

The treatment facility uses pretreatment, primary sedimentation, biological treatment, and disinfection treatment stages. The treatment processes are duplicated in case of failure or maintenance shutdowns. Ultraviolet disinfection is used instead of chlorine or other disinfectants. An anaerobic digester is used for treatment of the sludge. Frequent monitoring of performance focuses on conventional pollutants (BOD_5, $CBOD_5$, NH_3-N, TKN, pH, and TSS).

Before the sample collection and data analyses were performed, existing performance data were obtained from the WWTP to determine its treatability for conventional parameters during both wet and dry weather. Probability plots of influent and effluent TSS concentrations (Figure 8.2) and BOD_5 (Figure 8.3) show a consistent reduction of 2 to 2.5 logs, or 90+% removals for these constituents. Table 8.4 shows pH levels during some of the wet weather sampling dates. The pH levels were between 6.5 and 7.4, which are expected pH levels for wastewater treatment facilities. There were no observed differences in treatment performance during wet and dry weather for these conventional pollutants.

The design treatment flow rate was 91,000 m^3/day (24 million gallons per day, MGD) during Goodson's study, but the treatment plant averaged between 57,000 and 64,000 m^3/day (15 to 16.7 MGD) on dry weather days. Annual flows for the WWTP between 2008 and 2012 are shown in Table 8.5. The maximum daily flow rates have periodically exceeded the design flow rates. The NPDES permit for the Hilliard N. Fletcher WWTP lists an estimated 1.5 MGD (5678 m^3/day) storm water I&I that enters the treatment plant.

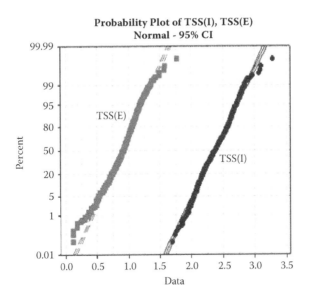

FIGURE 8.2
TSS concentration data from WWTP 2005–2008. (Reprinted with permission from Goodson, K., Treatability of Emerging Contaminants in Wastewater Treatment Plants during Wet Weather Flows, Doctoral Dissertation, University of Alabama, Tuscaloosa, 2013.)

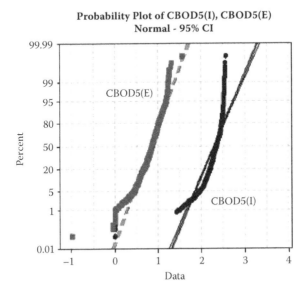

FIGURE 8.3
BOD_5 data from WWTP 2005–2008. (Reprinted with permission from Goodson, K., Treatability of Emerging Contaminants in Wastewater Treatment Plants during Wet Weather Flows, Doctoral Dissertation, University of Alabama, Tuscaloosa, 2013.)

TABLE 8.4

pH Concentrations for Sample Dates

Wet Weather Date	Influent pH	Effluent pH
January 16, 2010	7.23	6.80
March 2, 2010	7.42	6.79
April 24, 2010	6.94	6.61
June 25, 2010	6.93	6.62
October 24, 2010	7.02	6.72
November 2, 2010	6.92	6.48
March 9, 2011	7.24	6.64
September 20, 2011	7.09	7.04

Source: Goodson, K., Treatability of Emerging Contaminants in Wastewater Treatment Plants during Wet Weather Flows, Doctoral Dissertation, University of Alabama, Tuscaloosa, 2013.

TABLE 8.5

Average and Maximum Flows from Hilliard K. Fletcher WWTP

Flow Rate (m³/day)	2008	2009	2010	2011	2012
Annual average flow rate	61,700	57,900	57,900	62,800	59,000
Maximum daily flow rate	144,000	138,000	87,000	160,000	115,000

Source: NPDES permit no. AL0022713, 2010.

This permit was recertified in 2010. The drainage area serviced by the WWTP affects the potential storm water I&I.

8.2.2 Experimental Design

Goodson et al. (2012) collected eight sets of samples at the wastewater plant during wet weather and nine sets of samples during dry weather to compare concentrations and performance as a function of flow rates. Table 8.6 shows the number of samples obtained for different rain depth categories. The most significant changes in flow rates at the wastewater treatment facility occurred when rainfall depths were greater than 1.5 inches, but only two sets of samples were obtained during the largest rain category. Four coordinated composite samples were collected during each sampling day. Therefore, 24 samples were collected for PAHs, pharmaceuticals, and pesticides during wet weather samples, and 28 samples were collected during dry weather conditions.

The wet and dry conditions were the independent variables during the data analyses. Goodson designed the study to measure the effects of increased flows due to wet weather and changing concentrations on wastewater treatment of the PAHs (and other emerging contaminants). The dependent

TABLE 8.6

Ranges of Rainfall during Sampling

Rainfall Ranges (inches)	Sample Days for Tuscaloosa Treatment Plant
<0.1 (dry)	9
0.1–0.55 (wet)	3
0.56–1.0 (wet)	3
>1.0 (wet)	2

Source: Goodson, K., Treatability of Emerging Contaminants in Wastewater Treatment Plants during Wet Weather Flows, Doctoral Dissertation, University of Alabama, Tuscaloosa, 2013.

variables were the influent and effluent concentrations for each unit process. The wet weather samples were weather dependent and were therefore obtained as a judgmental sample design (when it was predicted to have moderate to large amounts of rainfall for the area). The dry weather samples were taken randomly, increasing variability. The sampling period began on January 16, 2010, and ended on November 12, 2012, therefore representing all seasons.

For each event, composite samples were obtained manually over a 6-hour period. Grab samples were taken at each sampling location as time composites over a 2-hour period, reflecting the movement of the water through the treatment plant. The samples were obtained at (1) the inlet, (2) the primary clarifier effluent, (3) the secondary clarifier effluent, and (4) after disinfection at the plant effluent. Each sample was obtained in 1 L prewashed amber glass bottles having Teflon-lined lids.

The samples were delivered to the chemistry laboratory at Miles College in Fairfield, Alabama, which used EPA Method 8310 for the PAH analyses. External calibrations used a blank and five concentrations for each analyte. These calibrations were verified by using internal standards. Method blanks were analyzed every 20 samples. The PAHs were extracted using methylene chloride in 2 L separation funnels. The extracts were condensed from 120 to 2 ml using Kuderna Danish (KD) glassware. The extracts were analyzed using gas chromatography with mass spectrometry (GC-MS) and selective ion methods (SIMs) for increased sensitivity for the targeted PAH compounds. The quality control objective for the laboratory blank was to obtain results in a concentration less than the specified detection limit. If the blank concentration was greater than the field samples, the values were rejected and the samples reanalyzed. The measured PAH detection limits ranged from 0.5 to 5 µg/L for the different compounds.

8.2.3 Results

Rainfall during wet weather was compared to the treatment plant flow rates to indicate the amount of increased flows associated with I&I. The flow rates

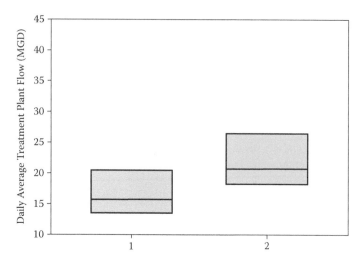

1: Treatment plant flows during dry weather.
2: Treatment plant flows during wet weather.

FIGURE 8.4
Average treatment flows during dry and wet weather. (Reprinted with permission from Goodson, K., Treatability of Emerging Contaminants in Wastewater Treatment Plants during Wet Weather Flows, Doctoral Dissertation, University of Alabama, Tuscaloosa, 2013.)

during dry weather conditions were also compared to flow rates during wet weather conditions, as shown in Figure 8.4. Most of the wet weather flows are larger than the dry weather flows, but there is some overlap. The Mann-Whitney rank sum test only identified a marginally statistically significant difference ($p = 0.07$), likely due to the small number of observations available that have both flow and rainfall data.

Table 8.7 includes calculations describing the amount of storm water I&I that could have affected the treatment plant for the different rain categories based on the observed rain and flow data. There is some uncertainty associated with these calculations, but they indicate that the amount of rainfall entering the sanitary sewer system as I&I and causing increased flows is likely very small (<2% even for the largest rains). However, the total sewage flow entering the treatment plant during large rains could be affected by large amounts of storm water I&I that entered the system by inflow (rapid entry) or infiltration (slower entry).

PAH concentrations were analyzed for samples collected at the influent and from each unit process's effluent at the treatment facility during both wet and dry conditions. Figures 8.5 and 8.6 are line plots for acenaphthene and phenanthrene, respectively, showing how the concentrations were reduced within the treatment facility. Acenaphthene indicated a steady decrease in concentration throughout the treatment facility, while phenanthrene indicated increased concentrations during the primary treatment stage and little

TABLE 8.7

Estimated Storm Water Infiltration and Inflow (I&I) for Different Rain Categories

Rain Range (mm)	Average Treatment Plant Flow (m³/day)	Increase over Base Treatment Plant Flow Assumed due to Storm Water I&I (m³/day)	Percentage of Total Treatment Plant Flow Associated with Storm Water I&I (%)	Estimated Storm Water I&I (m³/day/km²)[a]	Estimated Storm Water I&I (watershed mm)	Estimated Storm Water I&I as a Percentage of the Rain Depth (%)
0–2.5	67,000	0	0	0	0	0
5–12.5	68,000	1100	2	5.7	0.038	0.43
15–37	87,000	20,000	23	104	0.46	1.7
39–63	129,000	61,700	48	321	0.94	1.8

Source: Goodson, K., Treatability of Emerging Contaminants in Wastewater Treatment Plants during Wet Weather Flows, Doctoral Dissertation, University of Alabama, Tuscaloosa, 2013.

[a] Service area of 192 km².

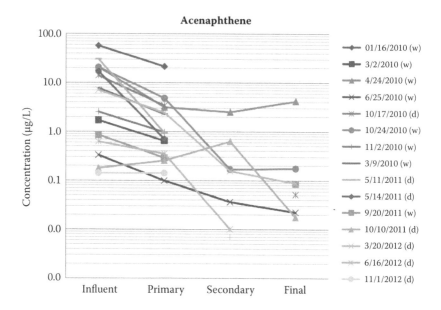

FIGURE 8.5
Concentration of acenaphthene for unit processes from Hilliard K. Fletcher WWTP. (Reprinted with permission from Goodson, K., Treatability of Emerging Contaminants in Wastewater Treatment Plants during Wet Weather Flows, Doctoral Dissertation, University of Alabama, Tuscaloosa, 2013.)

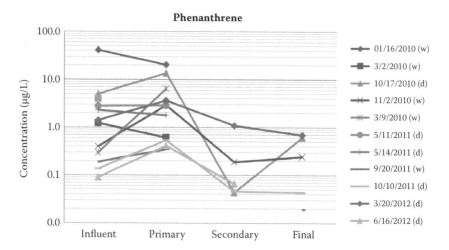

FIGURE 8.6
Concentration of phenanthrene for unit processes from Hilliard K. Fletcher WWTP. (Reprinted with permission from Goodson, K., Treatability of Emerging Contaminants in Wastewater Treatment Plants during Wet Weather Flows, Doctoral Dissertation, University of Alabama, Tuscaloosa, 2013.)

further treatment after disinfection, with most of the removal occurring during the secondary treatment stage. Comparing performance during wet and dry weather conditions did not indicate any apparent or statistically significant differences for the two weather conditions for any PAH except for naphthalene.

Since no statistical differences were identified between wet and dry periods, these data were combined for further analyses. Figure 8.7 shows box and whisker plots graphically contrasting the concentrations observed at each of the four sampling locations. These illustrate how the highly variable influent concentrations were reduced with treatment, resulting in much smaller final concentrations with smaller variabilities.

Naphthalene had the least removal reported in the literature, which was substantiated by the Tuscaloosa wet weather observations; dry weather naphthalene removals were significantly greater, as shown in Table 8.8. This table also shows that the overall PAH removals were high, ranging from 75 to 100% for all PAHs in this study, except for the wet weather naphthalene removals. Primary sedimentation was the most important removal process for most of the PAHs, while some had large removals during the secondary biological treatment process. The flow rate and corresponding weather conditions were not identified as significant factors.

Table 8.9 lists the calculated Mann-Whitney rank sum nonparametric comparison test results indicating the probability that the paired sample results are different. A p value of 0.05 or lower indicates at least a 95% confidence limit that they are different and is the traditional criterion for significance. Based on this value, no primary treatment or disinfection concentration differences were significant, while several were found to have statistically significant differences associated with the secondary treatment process. In most cases, the influent and effluent concentrations are significantly different, reflecting the combined benefit of all of the treatment processes, while a few are marginally significantly different (p between 0.05 and 0.1). Additional data would be needed to indicate statistically significant differences for wet vs. dry period values.

Goodson et al. (2012) also calculated the mass loads of PAHs entering the wastewater treatment facility during wet and dry periods. The mass loads were calculated using the average flow rates and concentrations for these rain conditions. Table 8.10 shows these results for three of the PAHs. The influent PAH mass loads indicate large mass increases during wet weather conditions. For these three compounds, acenaphthylene and acenaphthene had twice the daily mass discharges during wet weather compared to normal dry weather conditions. However, some PAHs indicated no apparent increases during wet weather increased flow conditions. The large variability in observed concentrations and the relatively few data observations obscured the smaller differences.

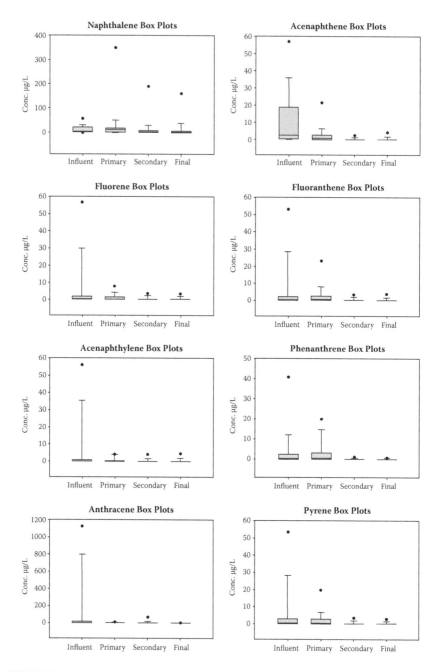

FIGURE 8.7
Box and whisker plots for PAH concentrations at sampling locations at Tuscaloosa WWTP (wet and dry weather data combined). (Reprinted with permission from Goodson, K., Treatability of Emerging Contaminants in Wastewater Treatment Plants during Wet Weather Flows, Doctoral Dissertation, University of Alabama, Tuscaloosa, 2013.)

TABLE 8.8

Most Important Removal Mechanisms and Overall Removals for PAHs

Constituent	Avg. Influent Concentration (μg/L)	Avg. Primary Effluent Concentration (μg/L)	Avg. Secondary Effluent Concentration (μg/L)	Avg. Concentration after UV (final effluent) (μg/L)	Avg. Overall Percentage Removal at ENH Wastewater Treatment Facility	Apparent Most Important Treatment Unit Process
Naphthalene (w)	15.3	4.74	25	22.7	-47	None
Naphthalene (d)	7.1	11.1	3.8	1.3	82	Secondary
Acenaphthene (w)	16.9	5.07	0.39	0.64	96	Primary
Acenaphthene (d)	7.70	0.82	0.10	0.02	99	Primary
Fluorene (w)	10.3	1.03	0.56	0.57	91	Primary
Fluorene (d)	0.67	1.19	0.04	0.05	93	Secondary
Fluoranthene (w)	10.3	4.23	0.54	0.53	95	Primary
Fluoranthene (d)	0.31	0.53	0.02	0.04	87	Secondary
Acenaphthylene (w)	10.5	0.60	0.61	0.67	92	Primary
Acenaphthylene (d)	0.08	0.58	0.01	0.02	75	Secondary
Phenanthrene (w)	6.14	4.36	0.05	0.15	98	Secondary
Phenanthrene (d)	1.56	0.77	0.16	0.12	90	Primary and secondary
Anthracene (w)	198	2.27	9.70	0.81	100	Primary
Anthracene (d)	60.07	0.18	0.24	0.15	100	Primary
Pyrene (w)	10.24	4.04	0.72	0.51	95	Primary and secondary
Pyrene (d)	0.66	0.95	0.13	0.13	80	Secondary

Source: Goodson, K., Treatability of Emerging Contaminants in Wastewater Treatment Plants during Wet Weather Flows, Doctoral Dissertation, University of Alabama, Tuscaloosa, 2013.

Note: w, wet; d, dry.

TABLE 8.9

Summary Mann-Whitney Comparison Statistical Test Results for PAHs (Combined Wet and Dry Weather Periods)

Constituent	Probability That Influent ≠ Primary Effluent	Probability That Primary Effluent ≠ Secondary Effluent	Probability That Secondary Effluent ≠ Final Effluent	Probability That Influent ≠ Final Effluent
Naphthalene	0.90	0.04	0.74	0.08
Acenaphthene	0.11	0.003	1.00	<0.001
Fluorene	0.65	0.04	0.70	0.04
Fluoranthene	0.83	0.011	0.75	0.04
Acenaphthylene	0.61	0.13	0.72	0.06
Phenanthrene	0.70	0.008	0.90	0.011
Anthracene	0.17	0.86	0.47	0.069
Pyrene	0.89	0.021	1.00	0.045

Source: Goodson, K., Treatability of Emerging Contaminants in Wastewater Treatment Plants during Wet Weather Flows, Doctoral Dissertation, University of Alabama, Tuscaloosa, 2013.

TABLE 8.10

Mass Loads of PAHs in Influent during Dry and Wet Weather

Average Mass Load	Phenanthrene (g/day)	Acenaphthene (g/day)	Acenaphthylene (g/day)
Dry	75	540	500
Wet	83	1100	1077

Source: Goodson, K., Treatability of Emerging Contaminants in Wastewater Treatment Plants during Wet Weather Flows, Doctoral Dissertation, University of Alabama, Tuscaloosa, 2013.

8.3 Conclusion

Chemical characteristics of PAHs indicate that they would mostly be associated with particulate solids and amenable to moderate to high levels of treatment at wastewater treatment facilities. Naphthalene was the only PAH compound studied whose chemical characteristics indicated poorer removal. Literature reports on PAH removals supported this finding, along with the observations obtained during this monitoring study during both wet and dry weather.

Storm water I&I was found to significantly affect the flow rate at the treatment facility during large rainfall events (>1.5 inches). There were also increases in the mass loads for many of the PAHs during wet weather.

However, the concentration changes through the treatment process did not support any statistically significant differences during wet weather compared to dry weather. Naphthalene did not show any removal during wet weather, while the other PAHs had 75 to 100% removals during both wet and dry weather. Additional monitoring, especially during large rains, would be needed to detect differences in PAH treatment during wet and dry weather conditions.

References

Andersson, T.A., K.M. Hartonen, and M.-L. Riekkola. (2005). Solubility of acenaphthene, anthracene, and pyrene in water at 50°C to 300°C. *Journal of Chemical Engineering Data* 50(4): 1177–1183.

ATSDR. (2011). Toxicological profile of polyaromic hydrocarbons. www.atsdr.gov (accessed April 12, 2012).

Blanchard, M., M.J. Teil, D. Ollivon, B. Garban, C. Chestérikoff, and M. Chevreuil. (2001). Origin and distribution of polyaromatic hydrocarbons and polychlorobiphenyls in urban effluents to wastewater treatment plants of the Paris area (France). *Water Research* 35(15): 3679–3687.

Goodson, K.L. (2013). Treatability of emerging contaminants in wastewater treatment plants during wet weather flows. Doctoral dissertation, University of Alabama, Tuscaloosa.

Goodson, K.L., R. Pitt, S. Subramaniam, and S. Clark. (2012). The effect of increased flows on the treatability of emerging contaminants at a wastewater treatment plant during rain events. *Proceedings of the Water Environment Federation* 2012(7): 7224–7237.

Mackay, D. (2001). *Multimedia environmental models: The fugacity approach*, 2nd ed. CRC Press, Boca Raton, FL.

Manoli, E., and C. Samara. (1999). Occurrence and mass balance of polycyclic aromatic hydrocarbons in the Thessaloniki sewage treatment plant. *Journal of Environmental Quality* 28(1): 176–186.

Pham, T.T., and S. Proulx. (1997). PCBs and PAHs in the Montreal Urban Community (Quebec, Canada) wastewater treatment plant and in the effluent plume in the St. Lawrence River. *Water Research* 31(8): 1887–1896.

Phillips, P.J., and A. Chalmers. (2009). Wastewater effluent, combined sewer overflows, and other sources of organic compounds to Lake Champlain. *Journal of the American Water Resources Association* 45(1): 45–57.

Streitwieser, A., C.H. Heathcock, and E.M. Kosower. (1998). *Introduction to organic chemistry*, 4th ed. Prentice-Hall, Upper Saddle River, NJ.

Tchobanoglous, G., F.L. Burton, and H.D. Stensel. (2003). *Wastewater engineering: Treatment and reuse* (Metcalf and Eddy, Inc.). McGraw-Hill, New York.

USEPA. (2008, January). Polycyclic aromatic hydrocarbons. EPA Fact Sheet. Washington, DC.

9

In Situ PAH Sensors

Woo Hyoung Lee,[1] Xuefei Guo,[2] Daoli Zhao,[2] Andres Campiglia,[3] Jared Church,[1] and Xiangmeng Ma[1]

[1]Department of Civil, Environmental, and Construction Engineering, University of Central Florida, Orlando, Florida

[2]Department of Chemistry, University of Cincinnati, Cincinnati, Ohio

[3]Department of Chemistry, University of Central Florida, Orlando, Florida

CONTENTS

9.1 Introduction

The detection of organic pollutants and their concentrations in soil, water, and air is essential for adequate environmental monitoring and analysis. There are recent needs for developing near-real-time environmental sensing and monitoring platforms for water quality with innovative algorithms to aid in natural/anthropogenic hazard responses (e.g., the *Deepwater Horizon* oil spill incident in 2010 and West Virginia chemical spill in 2014) and ecosystem restoration assessment. Among the many hydrophobic organic compounds, polycyclic aromatic hydrocarbons (PAHs) have been of great concern because of their carcinogenic and mutagenic properties— particularly four- to six-ring compounds [1, 2]. PAHs are neutral, nonpolar, and hydrophobic organic molecules comprised of two or more fused benzene rings. They are essentially insoluble and have very low vapor pressures; therefore, their measurement in aquatic environments is challenging, and the procedures for in situ sensors for PAHs have not yet been thoroughly explored.

Among the hundreds of PAHs present in the environment, the Environmental Protection Agency (EPA) lists 16 as "consent decree" priority pollutants. These are shown in Figure 9.1. For drinking waters, the EPA recommends the routine monitoring of benzo[a]pyrene. This is the most toxic PAH in the EPA list, and its individual concentration is often used as a measure of risk. According to the EPA, its maximum contaminant level (MCL) should not exceed 200 ng•L^{-1} [3]. The European Union and the World Health Organization (WHO) have regulated benzo[a]pyrene, fluoranthene, benzo[b]fluoranthene, benzo[k]fluoranthene, benzo[g,h,i] perylene, and indeno[1,2,3-c,d]pyrene. MCL values were set at 10 ng•L^{-1} for the highly toxic benzo[a]pyrene and 200 ng•L^{-1} for the remaining PAHs [4, 5]. Although EPA methodology provides reliable data on the environmental fate of EPA PAHs, the routine monitoring of numerous samples via fast, cost-effective, and environmentally friendly methods remains an analytical challenge.

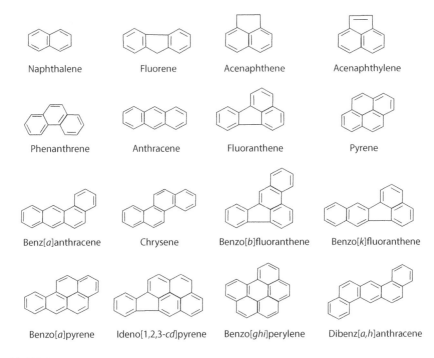

FIGURE 9.1
Molecular structures of the 16 EPA PAHs in order of carbon numbers and aromatic rings of PAHs.

9.2 Structure and Properties of PAHs for Analyses

9.2.1 Physical Structures

PAHs are comprised of two or more benzene rings bonded in linear, cluster, or angular arrangements (Figure 9.2). Although isomeric structures are similar, they have different thermodynamic stabilities. For example, both anthracene and phenanthrene have the same chemical formula ($C_{14}H_{10}$), but anthracene has approximately 6 kcal/mol (25 KJ/mol) of energy, which is less stable than phenanthrene. The reactivity and chemical stabilities of benzene (C_6H_6) are determined by the Hückel $4n + 2$ rule [6]; however, the Hückel rule does not fully explain PAHs aromatically, given that pyrene ($C_{16}H_{10}$) and coronene ($C_{24}H_{12}$) are aromatic but do not satisfy the Hückel rule. Inversely, PAH molecules such as annulene (C_nH_n or C_nH_{n+1}), which have localized π-electrons, are characterized as antiaromatic properties [7]. The Clar model seeks to explain chemical reactivity and other properties of PAHs by the extra stability of 6n π-electron benzenoid species [8]. In Clar's model, π-electrons that participate in aromatic

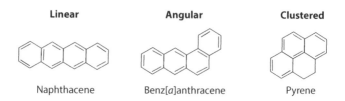

FIGURE 9.2
Examples of linear, angular, and clustered PAH molecular structures.

sextets (carbon $2p_z$ electrons) are assigned to particular rings in such a way as to obtain the maximum number of π-electron sextets (benzenoid). Hence, the Clar structure of PAHs is a resonance structure that has the maximum number of isolated and localized aromatic π-sextets by a minimum number of localized double bonds. With this rule, the PAHs with a higher number of π-sextets are more energetically stable than isomers with a lower number of aromatic π-sextets [9]. The external rings tend to be the most π-aromatic, and the internal rings tend to be sp^2 and sp– hybrid [10]. This inhomogeneity leads to the presence of different chemical reactivities between the peripheric and internal rings [11]. Clar's model can also be used to explain the decrease in stability and aromaticity of larger acenes. The higher reactivity of larger acenes that behave more like conjugated alkenes is consistent with the presence of a single aromatic sextet in the molecule (e.g., tetracene $[C_{18}H_{12}]$ and pentacene $[C_{22}H_{14}]$). There are more than 100 PAHs, but not all are of equal concern. The USEPA's National Waste Minimization Program (NWMP) has used the reporting criteria of the EPA's Toxic Release Inventory to identify 28 PAHs as priority pollutants (see Table 9.1) [12].

9.2.2 Electronic Structures

9.2.2.1 Electrochemical Properties

Electrochemical sensors encompass a large subclass of chemical sensors in which transduction is achieved by detecting some electrical properties of the targeting compounds on the electrode surface. The electrochemical methods take advantage of the coupling chemistry (in situ generation of oxidant) with electronic science (electron transfer), such as the electro-oxidation for PAH degradation. The oxidation and reduction chemistry of PAHs plays an important role in many sensors' design and application. In addition, the measurement of the oxidation/reduction for PAHs provides the experimental measurement for the various molecular orbitals. Table 9.2 shows a series of PAH reduction and oxidation potentials. The oxidation/reduction potentials are highly related to the length of conjugated rings and ring sizes. For example, indane shows the oxidation at +1.59 V vs. saturated calomel electrode (SCE) in acetonitrile, whereas the longer conjugated azulene exhibits the oxidation

TABLE 9.1

Names and CAS Numbers of 28 PAHs

Compound	CAS Number	Compound	CAS Number
Benz[a]anthracene	56–55–3	Dibenzo[a,l]pyrene	191–30–0
Benzo[a]phenanthrene (chrysene)	218–01–9	7H-Dibenzo[c,g]carbazole	194–59–2
Benzo[a]pyrene	50–32–8	7,12-Dimethylbenz[a]anthracene	57–97–6
Benzo[b]fluoranthene	205–99–2	Indeno[1,2,3-c,d]pyrene	193–39–5
Benzo[j]fluoranthene	205–82–3	3-Methylcholanthrene	56–49–5
Benzo[k]fluoranthene	207–08–9	5-Methylchrysene	3697–24–3
Benzo[j,k]fluorene (fluoranthene)	206–44–0	Acenaphthene	83–32–9
Benzo[r,s,t]pentaphene	189–55–9	Acenaphtylene	208–96–8
Dibenz[a,h]acridine	226–36–8	Anthracene	120–12–7
Dibenz[a,j]acridine	224–42–0	Benzo[g,h,i]perylene	191–24–2
Dibenzo[a,h]anthracene	53–70–3	Fluorene	86–73–7
Dibenzo[a,e]fluoranthene	5385–75–1	Phenanthrene	85–01–8
Dibenzo[a,e]pyrene	192–65–4	Pyrene	129–00–0
Dibenzo[a,h]pyrene	189–64–0	1-Nitropyrene	5522–43–0

Source: USEPA, Office of Solid Waste, Washington, DC, January 2008, http://www.epa.gov/osw/hazard/wastemin/priority.htm.

wave at +0.71 V [11, 13]. The oxidation of naphthalene (two conjugated rings) is at +1.54 V; the oxidation of anthracene, with three conjugated phenyl rings, shifts to +1.09 V; and the oxidation potential of naphthacene, with four conjugated phenyl rings, shifts further to +0.77 V [11, 13]. With the longer conjugated phenyl rings, the oxidation potentials shift to lower oxidation potential. It is also reported that the position of the conjugated rings affects the PAH electrochemistry. For example, for benz[a]pyrene the oxidation potential is +0.94 V, whereas for benz[e]pyrene it is +1.27 V under the same conditions [11]. The influences of number substitutions and substitution positions on the oxidation/reduction chemistry for PAHs were investigated, and it was found that the oxidation potentials shift to lower oxidation potentials with the introduction of the electron donor groups [14, 15].

9.2.2.2 Absorption Spectroscopy

Figure 9.3 shows the different electron transitions and fates of the excited PAHs after light absorption. PAHs exhibit absorption spectrum with vibronic structure due to the oscillations of the carbon skeleton. The absorption spectra of symmetric PAHs can be determined by the transitions from the ^1A state (ground state of the system) to the ^1L_a, ^1L_b, ^1B_a, and ^1B_b states (superscript 1 refers to singlet state). For example, the three long-wavelength transitions observed in benzene and its derivatives correspond to the excitation of ^1L (^1L_a and ^1L_b) and ^1B (^1B_a and ^1B_b) states, which in benzene are degenerate.

TABLE 9.2

Oxidation/Reduction Potentials for PAHs

No.	Compound	$E_{1/2}$ (oxidation)[a]	$E_{1/2}$ (reduction)[a]	Reference
1	Indane	1.59		[11]
2	Naphthalene	1.54	−2.49	[13]
3	1-Methylnaphthalene	1.43		[13]
4	2-Methylnaphthalene	1.45		[13]
5	2,3-Dimethylnaphthalene	1.35		[13]
6	2,6-Dimethylnaphthalene	1.36		[13]
7	1-Methoxynaphthalene	1.38	−2.65	[15]
8	2-Methoxynaphthalene	1.52	−2.60	[15]
9	1,3-Dimethoxynaphthalene	1.26	−2.61	[14]
10	1,4-Dimethoxynaphthalene	1.10	−2.69	[15]
11	1,5-Dimethoxynaphthalene	1.28	−2.75	[14]
12	1,6-Dimethoxynaphthalene	1.28	−2.68	[15]
13	1,7-Dimethoxynaphthalene	1.28	−2.67	[15]
14	1,8-Dimethoxynaphthalene	1.17	−2.72	[15]
15	2,3-Dimethoxynaphthalene	1.39	−2.73	[15]
16	2,6-Dimethoxynaphthalene	1.33	−2.60	[15]
17	2,7-Dimethoxynaphthalene	1.47	−2.68	[15]
18	1,4,5,8-Tetramethoxynaphthalene	0.70	−2.69	[15]
19	1-Dimethylaminonaphthalene	0.75	−2.58	[15]
20	1,5-bis(Dimethylamino)naphthalene	0.58	−2.64	[14]
21	2-Dimethylaminonaphthalene	0.67	−2.63	[14]
22	2,6-bis(Dimethylamino)naphthalene	0.26	−2.71	[15]
23	2,7-bis(Dimethylamino)naphthalene	0.57	−2.77	[15]
24	1-(Methylthio)naphthalene	1.32	−2.25	[15]
25	2-(Methylthio)naphthalene	1.36	−2.28	[15]
26	1,4-bis(Methylthio)naphthalene	1.07	−2.10	[15]
27	1,5-bis(Methylthio)naphthalene	1.26	−2.15	[14]
28	1,8-bis(Methylthio)naphthalene	1.09	−2.22	[15]
29	2,3-bis(Methylthio)naphthalene	1.35	−2.21	[14]
30	2,6-bis(Methylthio)naphthalene	1.10	−2.24	[15]
31	2,7-bis(Methylthio)naphthalene	1.33	−2.25	[15]
32	1,4,5,8-Tetraphenylnaphthalene	1.39	−1.98	[15]
33	1,5-Dimethoxy-4,8-bis(methylthio) naphthalene	0.70	−2.42	[15]
34	1,5-Dimethoxy-4,8-diphenoxnaphthalene	0.98	−2.47	[15]
35	1-Naphthylamine	0.54		[13]
36	2-Naphthylamine	0.64		[13]
37	Azulene	0.71		[13]
38	Biphenyl		−2.55	[11]
39	4-Methoxybiphenyl	1.53	−2.73	[15]

(Continued)

TABLE 9.2 *(Continued)*

Oxidation/Reduction Potentials for PAHs

No.	Compound	$E_{1/2}$ (oxidation)[a]	$E_{1/2}$ (reduction)[a]	Reference
40	4,4′-Dimethoxybiphenyl	1.30		[15]
41	3,3′-Dimethoxybiphenyl	1.60	−2.54	[15]
42	2,2′-Dimethoxybiphenyl	1.51		[15]
43	4,4′-bis(Methylthio)biphenyl	1.25		[14]
44	3,3′-bis(Methylthio)biphenyl	1.47		[14]
45	Diphenylmethane			[11]
46	trans-stilbene	1.43	−2.08	[11]
47	Acetnaphthene	1.21	−2.67	[11]
48	Acenaphthylene	1.21		[11]
49	Fluorene			[11]
50	2-Fluorenamine	0.44		[13]
51	Phenanthrene	1.5	−2.44	[11]
52	Anthracene	1.09	−1.95	[11]
53	9-Methylanthracene	0.96		[13]
54	2-Aminoanthracene	0.44		[13]
55	9,10-Dimethylanthracene	0.87		[13]
56	9-Methoxyanthracene	1.05	−1.92	[15]
57	9,10-Dimethoxyanthracene	0.98	−1.9	[15]
58	9,10-bis(Methylthio)anthracene	1.11	−1.55	[15]
59	9,10-bis(2,6-Dimethoxyphenyl)anthracene	1.18	2.08	[15]
60	9,10-bis(Phenylethyny1)anthracene	1.16	−1.29	[14]
61	9,10-Diphenoxyanthracene	1.2	−1.71	[15]
62	Pyrene	1.16	−2.09	[11]
63	1,6-bis(Dimethylamino)pyrene	0.49	−2.16	[15]
64	1,6-Dimethoxypyrene	0.82	−2.19	[15]
65	1,6-bis(Methylthio)pyrene	0.96	−1.83	[15]
66	Fluoranthene	1.45	−1.74	[11]
67	Chrysene	1.35	−2.25	[11]
68	Triphenylene	1.55	−2.46	[11]
69	p-Terphenyl			[11]
70	Naphthacene	0.77	−1.58	[11]
71	3-Methylcholanthrene	0.87		[13]
72	Benz[a]anthracene	1.18		[11]
73	Benzo[a]pyrene	0.94	−2.1	[11]
74	Benzo[e]pyrene	1.27	−2.13	[11]
75	3,4-Benzpyrene	0.94		[13]
76	Perylene	0.85	−1.67	[11]
77	7,12-Dimethylbenz[a]anthracene	0.96		[11]
78	3-Methylcholanthene	0.87		[11]
79	Benzo[g,h,i]perylene	1.01		[11]

(Continued)

TABLE 9.2 *(Continued)*

Oxidation/Reduction Potentials for PAHs

No.	Compound	$E_{1/2}$ (oxidation)[a]	$E_{1/2}$ (reduction)[a]	Reference
80	Dibenz[a,c]anthracene	1.25		[11]
81	Dibenz[a,h]anthracene	1.19		[11]
82	Dibenz[a,j]anthracene	1.26		[11]
83	1,2-Benzanthracene	1.18		[13]
84	1,2,5,6-Dibenzanthracene	1.19		[13]
85	1,2,3,4-Dibenzanthracene	1.25		[13]
86	1,2,7,8-Dibenzanthracene	1.26		[13]
87	1-Methyl-1,2-benzanthracene	1.14		[13]
88	2-Methyl-1,2-benzanthracene	1.14		[13]
89	3-Methyl-1,2-benzanthracene	1.14		[13]
90	4-Methyl-1,2-benzanthracene	1.15		[13]
91	5-Methyl-1,2-benzanthracene	1.15		[13]
92	6-Methyl-1,2-benzanthracene	1.15		[13]
93	7-Methyl-1,2-benzanthracene	1.08		[13]
94	8-Methyl-1,2-benzanthracene	1.13		[13]
95	9-Methyl-1,2-benzanthracene	1.15		[13]
96	10-Methyl-1,2-benzanthracene	1.14		[13]
97	11-Methyl-1,2-benzanthracene	1.14		[13]
98	12-Methyl-1,2-benzanthracene	1.07		[13]
99	Picene	1.33		[13]
100	1,2-Benzpyrene	1.27		[13]
101	3,4-Benztetraphene	1.01		[13]
102	1,12-Benzoperylene	1.01		[13]
103	1,2,4,5-Dibenzpyrene	1.01		[13]
104	Coronene	1.23	−2.07	[11]

[a] Measured vs. saturated calomel electrode.

The λ_{max} of PAH absorption spectra increases as the size of the conjugated macrocyclic increases (i.e., with more conjugated bonds) [16]. For example, naphthalene shows a strong absorption band at 220 nm, whereas anthracene, triphenylene, and pyrene show their absorption maxima at 251, 256, and 239 nm, respectively. Pyrene shows a more complicated spectrum with strong absorption bands at 239, 272, and 334 nm. The absorption spectra of all these derivatives show shoulders and subsidiary peaks on the short wavelength side of each of the strong absorption bands. These subsidiary peaks derive from vibronic Frank-Condon absorption components [16]. The shifting of λ_{max} toward longer wavelengths is due to the frequency of $S_1 \rightarrow S_0$ transition decreases (Figure 9.3). This increase in π-conjugation causes the energy spacing between the highest occupied molecular orbital (HOMO) and lowest unoccupied molecular orbital (LUMO) to decrease, resulting in

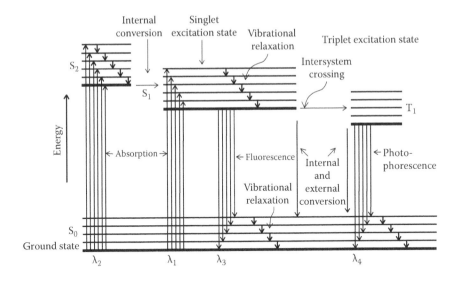

FIGURE 9.3
Electron transition energy levels. (Reprinted with permission from Okparanma and Mouazen, *Applied Spectroscopy Reviews*, 48, 458–486, 2013. [17])

a red shifting (increasing in wavelength) of the absorption spectrum; e.g., tetracene is yellow, while anthracene is white. The decrease in transition frequency is less pronounced for angular benzenoids (phenes). This is reflected in the observed color of pentacene (yellow) compared to its angular isomer pentaphene (white) [11].

Substitution on the PAHs can be used to tune the absorption spectra for a red/blue shift. For example, the introductions of methyl, phenyl, and vinyl groups into the aromatic rings lead to the red shift of the absorption spectrum. In addition, the heteroatoms (e.g., nitrogen, oxygen, and sulfur) introduced into the aromatic ring make the spectrum red shifts.

9.2.2.3 Fluorescence Spectroscopy

As depicted in the Jablonski diagram in Figure 9.3, fluorescence involves the emission of a photon with a quantum energy ($E = h\nu$) that corresponds to the energy difference between the lowest vibrational level of the S_1 state and one of the vibrational levels of the electronic ground state. Figure 9.3 suggests that a fluorescence emission measurement would yield a spectrum that consists of a number of discrete lines, but that is usually not observed in molecular systems owing to several spectral broadening processes; in conventional fluorescence spectra of PAHs, band widths are typically 200 to 600 cm^{-1}.

The fluorescence spectrum is highly related to the absorption spectrum. In general, the fluorescence spectrum is the strong mirror image of the absorption spectrum, varying only in intensity of ratios, probably due to different quantum yields of the molecular excited state. It reflects energy changes that accompany the return of a π-electron from the lowest vibrational level of the excited state to various vibrational levels of the ground state. Such a transition of a π-electron that remains in the excitation state for about 10^{-8} s gives rise to a set of fluorescence bands. The fluorescence spectroscopy provides information on the ground and excited electronic states, Franck-Condon factors, changes in geometry, mixing of excitation states, and dynamics in the excited states. The constant spacing between the fluorescence bands corresponds to the energy differences of the vibrational levels of the ground state, whereas the spacing of bands in the absorption spectra corresponds to the vibrational energy levels of the excited state [18].

In homogenous media, PAHs are highly fluorescent. However, the low values of fluorescence quantum yields for some PAHs are due to competing $S_1 \rightarrow T_1$ intersystem crossing (Figure 9.3) [11]. The 0–0 transition, which is absorption and emission between lowest vibrational levels of the molecules' ground and excited states, exists in the absorption and emission spectroscopy. The 0–0 transition band is normally less intense than other bands in the emission spectra due to the π-π* transition. In the case of $^1S \rightarrow S_0$ allowed transition, the intensity is stronger, such as in the case of anthracene. Planar rigid PAHs emit more efficiently than nonplanar counterparts, due to their less efficient radiationless deactivation of the excited singlet state. Fluorescence spectra and lifetimes of PAHs are sensitive to solvent polarity. Solvent-induced fluorescence spectral changes can be rationalized qualitatively in a relatively straightforward manner [19]. Excitation promotes the PAH solute from a ground state of low dipole moment to one of the vibrational levels of the first electronic excited state, S_v^*, with an accompanying electron distribution in the surrounding solvent molecules. Insufficient time exists, however, for solvational sphere molecules to physically reorient with the new PAH dipole moment. Relaxation from the vibrationally excited S_v^* level to the excited S_0^* level occurs whenever solvent molecules rotationally reorient to a more stable dipole configuration during the excited state's lifetime. Emission of the fluorescence photon returns the PAH molecule to the ground S_v state and solvational molecules to their initial electronic configuration. Subsequent rotation of solvent molecules to the ground-state dipole orientation restores the system to its original state. Transition probabilities and energy separations between the different energy levels vary with each solute-solvent pair, and give rise to observed intensity ratio changes and emission wavelength shifts [20, 21]. For example, the most intense peak of emission spectra of benzo[k]fluoranthene shifts in different solvents from nonpolar *n*-hexadecane hydrocarbon to moderately polar dichloromethane to very polar dimethyl sulfoxide (Figure 9.4) [22]. The emission intensity band ratios for benzo[k]fluoranthene (I/II) ranged from 1.48 in cyclohexane to 1.04 in dichloromethane and 1.0 in dimethyl sulfoxide (DMSO) [22] (Table 9.3).

Polycyclic aromatic nitrogen heterocycles (PANHs) are susceptible to protonation, particularly in acid environments. Protonation of the nitrogen

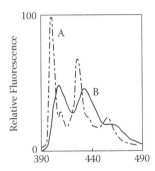
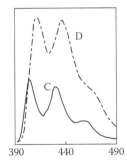
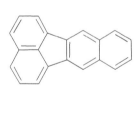

FIGURE 9.4

Fluorescence emission spectra of benzo[k]fluoranthene (structure is at right) dissolved in [A (-•-•-)] *n*-hexadecane, [B (— —)] dichloromethane, [C (— -)] butyl acetate, and [D (-•-•-)] dimethyl sulfoxide. In butyl acetate, emission bands occur at 405, 430, and 458 nm. (From Tucker et al., *Applied Spectroscopy*, 45, 1699–1705, 1991.)

TABLE 9.3

Ratios of Emission Intensities for Benzo[k]fluoranthene in Selected Organic Solvents

Solvent	I/II Ratio	Solvent	I/II Ratio
Cyclohexane	1.48	Chloroform	1.10
2,2,4-Trimethylpentane	1.46	Butyl acetate	1.11
n-Hexadecane	1.43	Methanol	1.08
Carbon tetrachloride	1.31	Dichloromethane	1.04
Dibutyl ether	1.24	Acetonitrile	1.04
Benzene	1.13	N,N-Dimethyl formamide	1.00
2-Propanol	1.10	Dimethyl sulfoxide	1.00

Source: Tucker et al., *Applied Spectroscopy*, 45, 1699–1705, 1991.

lone electron pair by a hydrogen ion often results in the loss of emission fine structure accompanied by a sizable red shift in emission spectrum. The degree of protonation should be reflected by solvent acidity and PANH basicity. Emission spectroscopy of nonprotonated diphenanthro [9,10,1def; 1′,10′,9′hij]phthalazine (DPP) shows that it is highly structured, with the maximum peaks at 450, 475, and 520 nm. The spacing is related to C=C or C=N stretching modes. After titration with acid, the emission spectrum loses the structure information, with only one peak maximum at 514 nm (Figure 9.5). In addition, maximum emission peaks shift to the longer wavelength [19].

In heterogeneous media, fluorescence of PAH experiences a broadening effect when adsorbed on solid surfaces (e.g., adsorbed to silica). Dabestani et al. [23–30] used steady-state fluorescence techniques to probe changes in the emission of PAHs adsorbed on silica or alumina. These studies revealed that the increased emission of PAHs with the higher concentration was attributed

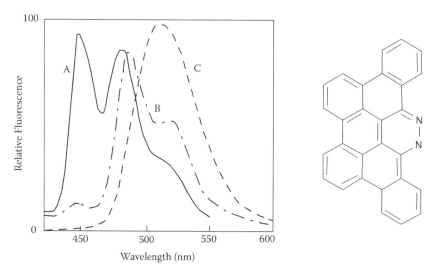

FIGURE 9.5

Fluorescence emission spectra of the neutral (A, in dimethyl sulfoxide) and protonated (C, in HC104-trifluoroethanol) forms of DPP (structure at right). Curve B was recorded in trifluoro-ethanol and shows only partial protonation. Protonation of the nitrogen heteroatom results in loss of emission fine structure accompanied by a red shift in emission wavelengths. (From Tucker et al., *Applied Spectroscopy*, 46, 229–235, 1992.)

to the formation of ground-state pairs and/or aggregations. The emissions of the aggregations exhibited either a structured emission (e.g., anthracenes and tetracene) or a structureless emission (excimer-like) (e.g., acenaphthene, fluorene, naphthalene, 1-methoxynaphthalene, and pyrene). The observed change in the emission of phenanthrene at high surface coverages is due to microcrystal formation [24]. Also, this change is attributed to the formation of pyrene excimer on silica to ground-state pairing [31].

9.3 Analytical Methods of PAHs in Laboratory Settings

EPA methodology for the analysis of PAHs in water samples follows the classical pattern of sample extraction and chromatographic analysis. The method of choice for PAH extraction and preconcentration is solid-phase extraction (SPE). PAH determination is carried out via high-performance liquid chromatography (HPLC) or gas chromatography–mass spectrometry (GC-MS). When HPLC is applied to highly complex samples, EPA recommends the use of GC-MS to verify compound identification and to check peak purity of HPLC fractions [32, 33].

9.3.1 Solid-Phase Extraction

The main mechanism of extraction is a nonpolar interaction (of the van der Waals type) between the PAH and a nonpolar sorbent such as an octadecyl (C_{18}) bonded-phase material. The experimental procedure for SPE consists of four steps: conditioning, retention, rinsing, and drying. The solid sorbent must undergo proper conditioning to wet the packing material before the sample passes through the SPE device. The conditioning step "activates" the sorbent by treatment with an organic solvent, typically methanol. Under these conditions, the bonded-organic moiety is more open and available for interaction with the solute. After conditioning, the sample is loaded into the SPE device by gravity feed, pumping, or vacuum. During this step, PAHs are extracted and preconcentrated in the solid sorbent. Some of the matrix concomitants may also be retained along with PAHs, making matrix composition more challenging for further chromatographic analysis. A rinsing step is then used to remove potential interferences. For aqueous samples, a water-organic-solvent mixture is usually used as the rinsing solvent. The final step is the elution of PAHs from the sorbent material with an appropriate solvent specifically chosen to disrupt PAH-sorbent interactions.

9.3.2 High-Performance Liquid Chromatography

According to EPA Method 550.1, the HPLC analysis of the 16 EPA PAHs should use a combination of ultraviolet-visible (UV-VIS) absorption and room temperature fluorescence (RTF) detection. UV-VIS should be used for the detection of naphthalene, acenaphthylene, acenaphthene, and fluorene. The determination of the remaining PAHs should be based on RTF detection. In both cases, PAH identification is solely based on the comparison of retention times recorded from samples and standards. Separation is carried out on a Supelco LC-PAH column with the following characteristics: 25 cm length, 4 mm diameter, and 5 μm average particle diameters. The settings and conditions of analysis include 2.0 ml min^{-1} flow rate, isocratic elution with 65/35 acetonitrile/water for 2 min, and then linear gradient to 100% acetonitrile over 22 min. The total separation time is approximately 30 min while consuming 60 ml of mobile phase (acetonitrile-water) per sample. Considering the gradient elution of chromatographic separation, this mobile phase volume is equivalent to 42 ml of acetonitrile. Numerous reported methods use methanol as the organic solvent in the mobile phase [34–37]. Separation of the 16 EPA PAHs takes 53 min and consumes approximately 66 ml of methanol per sample. The main reason methanol is often preferred over acetonitrile is its lower cost.

Figure 9.6 shows a typical HPLC chromatogram of 16 EPA PAHs in *n*-octane.

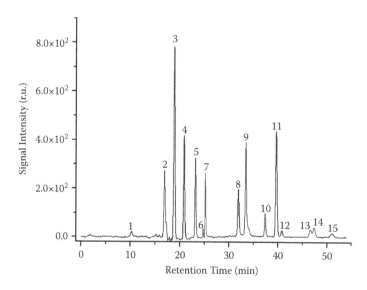

FIGURE 9.6

Typical HPLC-RTF chromatogram of a 25-ng•ml⁻¹ standard of 16 EPA PAHs in *n*-octane. The peaks are labeled as follows: (1) naphthalene, (2) acenaphthene, (3) fluorene, (4) phenanthrene, (5) anthracene, (6) fluoranthene, (7) pyrene, (8) benz[a]anthracene, (9) chrysene, (10) benzo[b] fluoranthene, (11) benzo[k]fluoranthene, (12) benzo[a]pyrene, (13) dibenz[a,h]anthracene, (14) benzo[g,h,i]perylene, and (15) indeno[1,2,3-c,d]pyrene. Acenaphthylene does not appear in the chromatogram because it shows no fluorescence. (Data collected in Dr. Campiglia's lab.)

9.3.3 Gas Chromatography–Mass Spectrometry

When HPLC is applied to highly complex samples and numerous unfamiliar peaks appear in the chromatogram, the use of a supporting analytical technique is recommended to verify compound identification and to check peak purity of HPLC fractions. The EPA method of choice (Method 525.1) is based on GC-MS. Separation of the 16 EPA PAHs is carried out with a 5% phenyl methyl siloxane stationary phase (0.25 μm film thickness) packed in a 30-m-length column with a 0.25-mm internal diameter. Ultra-pure helium carrier gas is used as the mobile phase (flow rate of 33 cm·s⁻¹) with either a single- (160–320°C at 6°C min⁻¹) or a multiramp (130–180°C at 12°C min⁻¹, 180–240°C at 7°C·min⁻¹, and 240–320°C at 12°C·min⁻¹) temperature oven program. The total separation time is approximately 32 and 25 min for the single- and multiramp temperature programs, respectively. PAH identification is based on retention times and mass spectra. For positive identification, the retention time of the unknown peak should be within 10 s of the reference standard retention time. All ions that are present above 10% relative abundance in the mass spectrum of the standard should also be present in the mass spectrum of the unknown within an agreement of 20%.

Figure 9.7 shows GC-MS chromatogram of a standard mixture of the 16 EPA PAHs in *n*-octane.

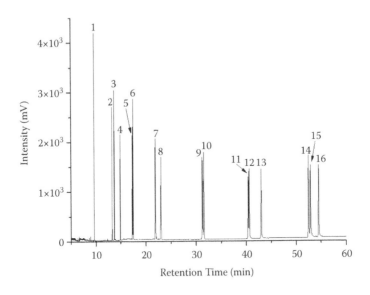

FIGURE 9.7
GC-MS chromatogram of a standard mixture of the 16 EPA PAHs at the 10-ng ml^{-1} concentration level in *n*-octane. The peaks are labeled as follows: (1) naphthalene, (2) acenaphthylene, (3) acenaphthene, (4) fluorine, (5) phenanthrene, (6) anthracene, (7) fluoranthene, (8) pyrene, (9) benz[a]anthracene, (10) chrysene, (11) benzo[b]fluoranthene, (12) benzo[k]fluoranthene, (13) benzo[a]pyrene, (14) indeno[1,2,3-c,d]pyrene, (15) dibenz[a,h]anthracene, and (16) benzo[g,h,i] perylene. (Data collected in Dr. Campiglia's lab.)

Although EPA methodology provides reliable data on the environmental fate of EPA PAHs, the routine monitoring of numerous samples via fast, cost-effective, and environmentally friendly methods remains an analytical challenge. One of the challenges is the time consumed; typically, 1 L of water is processed through the SPE device in approximately 1 h. The rather large water volume and long sample processing time are recommended to reach detectable concentrations and quantitative removal of PAHs from water samples. Chromatographic elution times of 30 to 60 min are typical, and standards must be run periodically to verify retention times. If concentrations of targeted PAHs are found to lie outside the detector's response range, the sample must be diluted (or concentrated), and the process repeated.

9.3.4 Laser-Excited Time-Resolved Shpol'skii Spectroscopy

A powerful technique for the direct determination of PAHs—no chromatographic separation—in environmental samples is Shpol'skii spectroscopy [38]. This is a fluorescence technique using a dilute solution of a guest molecule (PAH) in a solvent host (usually an *n*-alkane), where the solvent freezes to 77 K or below into an ordered polycrystalline matrix.

If the dimensions of the PAH and solvent match up well enough, PAH molecules occupy a small number of crystallographic sites in the host matrix. Matrix isolation of guest molecules reduces inhomogeneous band broadening. The combination of reduced thermal and inhomogeneous broadening produces vibrationally resolved spectra with sharp line widths with tremendous potential for qualitative and quantitative determination. Higher spectral resolution can be obtained by site-selective excitation, which refers to the excitation of guest molecules occupying one type of crystallographic site. PAHs occupying the same crystallographic site produce identical excitation and emission spectra. PAHs occupying different sites produce identical spectral profiles, which are slightly shifted by small wavelength increases (typically less than 1000 cm^{-1}). Multisite excitation produces spectra with contributions from PAH molecules in all crystallographic sites. The simplest spectra, and therefore the narrowest full width at half maxima, are obtained by site-selective excitation, which is best accomplished with narrow-band laser sources.

Particularly attractive for cryogenic measurements at liquid nitrogen (77 K) and liquid helium (4.2 K) temperatures is the use of fiber optic probes [39, 40]. The one excitation and the six collection fibers are fed into a section of copper tubing that provides mechanical support for lowering the probe into the liquid cryogen. At the analysis end, the excitation and emission fibers are bundled with vacuum epoxy and fed into a metal sleeve for mechanical support. The copper tubing is flared, stopping a swage nut tapped to allow for the threading of the sample vial. At the instrument end, the emission fibers are bundled with vacuum epoxy in a slit configuration, fed into a metal sleeve, and aligned with the entrance slit of the spectrometer. The tip of the probe is positioned above the solution surface as the sample tube is lowered into a container filled with liquid cryogen. The cell is allowed to cool for 90 s prior to fluorescence measurements, to ensure complete sample freezing. Samples are frozen in a matter of seconds.

Cryogenic probes can be easily coupled to both narrow-band excitation sources (lasers) [39, 40] and broad-band Xenon lamps usually available in commercial spectrofluorimeters [41]. Pulsed laser excitation sources offer prospects for time-resolving fluorescence spectra. Adding the temporal dimension to the highly resolved Shpol'skii spectra provides a powerful tool for the determination of EPA PAHs in complex samples. Campiglia and coworkers [42] have shown that the combination of a pulsed tunable dye laser, a pulsed delay generator, a spectrograph, and an intensifier-charged coupled device is well suited for the rapid collection of wavelength time matrices (WTMs) and time-resolved excitation-emission matrices (TREEMs) in the fluorescence (ns to µs) and phosphorescence (ms) time domains. WTMs consist of series of emission spectra recorded at different time delays from the laser excitation pulse. TREEMs are basically series of excitation-emission matrices recorded at different delay and gate times from the laser excitation pulse. The complete TREEM data set consists of emission intensity as a

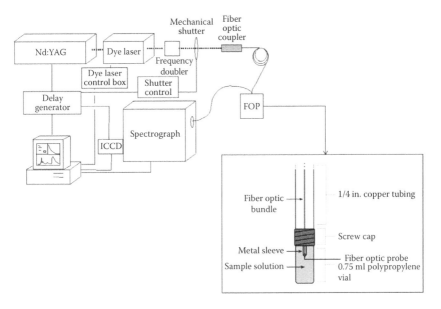

FIGURE 9.8
Instrumentation for laser-excited time-resolved Shpol'skii spectroscopy. FOP = fiber optic probe, ICCD = intensified charge coupled device. Not to scale.

function of excitation wavelength, emission wavelength, and delay and gate times after the short duration of the excitation pulse.

Figure 9.8 shows an instrumentation configuration for laser-excited time-resolved Shpol'skii spectroscopy (LETRSS).

The use of cryogenic fiber optic probes facilitates the hyphenation of laser-excited time-resolved Shpol'skii spectroscopy to sample preconcentration techniques for the analysis of aqueous samples liquid-liquid extraction (LLE) or solid-phase extraction (SPE) with laser-excited time-resolved Shpol'skii spectroscopy (LLE-LETRSS, SPE-LETRSS), and solid-phase nanoextraction (SPNE) methods have been developed for the determination of PAHs in both HPLC fractions and sample extracts with chromatographic separation [43–51]. SPNE refers to the extraction of PAHs with gold nanoparticles. According to their 77 K standard fluorescence lifetimes, EPA PAHs can be categorized in three distinct groups: PAHs with relatively short ($\tau \leq 11.1$ ns), medium (39.9 ns $\leq \tau \leq 61.7$ ns), and long (182.8 ns $\leq \tau \leq 523.1$ ns) lifetimes. By selecting the appropriate time window (i.e., delay and gate times) during the total fluorescence decay of the sample, it is possible to minimize spectral overlapping and determine each PAH without previous chromatographic separation. Spectral purities at target wavelengths are monitored via fluorescence decays. Single exponential decays with statistically equivalent lifetimes to the pure standards confirm peak assignment to the correct PAH.

Figure 9.9 shows a 77 K fluorescence spectrum of a river water sample, spiked with 15 EPA PAHs.

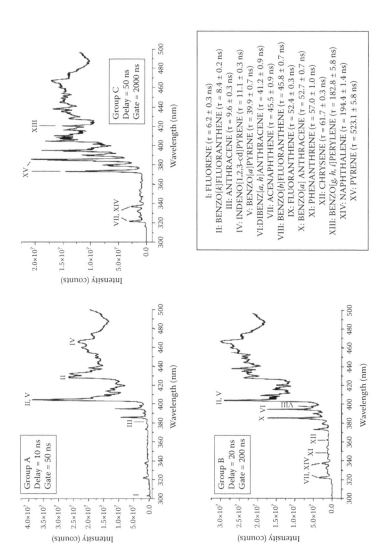

FIGURE 9.9

A 77 K fluorescence spectrum of a river water sample of unknown composition spiked with 15 EPA PAHs at the ng.ml⁻¹ concentration level. Shpol'skii spectra were recorded using optimum delay and gate times for each PAH group. Lifetimes and standard deviations based upon six replicate measurements from different frozen aliquots of standard solutions. Lifetime measurements were made at the maximum emission wavelength of each PAH. (Data collected in Dr. Campiglia's lab.)

9.4 In Situ PAH Sensors and Their Application for In Situ Monitoring

As was discussed in previous sections, conventional methods of detecting PAHs in the environment include liquid-liquid extraction (LLE) [52], solid-phase extraction (SPE) [53–56], supercritical fluid extraction (SFE) [57] in combination with gas chromatography (GC) [55, 58], high-performance liquid chromatography (HPLC) [56] coupled with mass spectrometry (MS) [55], and flame ionization [58]. These methods are accurate; however, they are laborious, time-consuming, requiring costly analytical resources, and generally performed in a laboratory. To monitor environments with high temporal and spatial variations, in situ methods are desired. There are several current techniques for in situ monitoring of PAHs as well as many cutting-edge sensing methods that show in situ detection potential. This section will discuss various approaches to the environmental monitoring of PAH.

9.4.1 In Situ Fluorometer

Fluorescence is the emission of light by a molecule whose electron has been excited to a higher energy level by an absorption of energy (i.e., a photon). As the electron returns to its ground state, energy is lost by collision, nonradiative decay, and other processes, leaving the energy of the emitted photon with lower energy than the excitation energy. This loss of energy is called Stokes shift and is the fundamental basis by which fluoroscopy operates [59]. PAHs provide inherence fluorescent properties in the UV-VIS spectral domain (200–400 nm) because of their aromatic structure [60]. The energy sharing and unpaired electron structure of the carbon ring allow for absorption and fluorescence of photons.

Figure 9.10 shows the general layout of a submersible fluorometer.

Fluorescence spectroscopy, especially the use of excitation-emission matrices (EEMs), has been a useful method for analyzing PAHs in aquatic environments. The process is typically time-consuming and performed ex situ, but with the recent development of submersible fluorometers, PAH measurements can now be obtained in situ and in real time [61]. The EnviroFlu-HC submersible fluorometer (TriOS Optical Sensors, Rastede, Germany) is a commercially available fluorometer designed for the detection of PAHs. It uses a xenon lamp to provide excitation light at 254 nm with a full width at half maximum (FWHM) of 25 nm (~254 ± 12.5 nm) and detects emission light at 360 nm with a FWHM of 50 nm (~360 ± 25 nm) [60]. These wavelengths allow for quantification of PAHs, in particular phenanthrene. Figure 9.10 shows the general optical layout of submersible fluorometers. A limitation to fluoroscopic sensing is that other aromatic molecules (e.g., humic substances) will interfere with the quantification of PAHs [60]. Applications of submersible fluorometers include environmental, drinking water, and

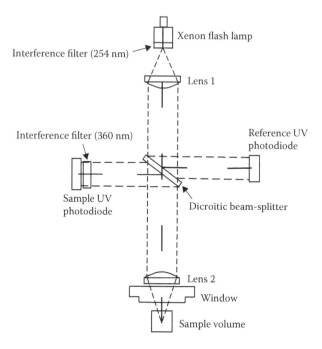

FIGURE 9.10
General layout of submersible fluorometer. (From Tedetti et al., *Marine Pollution Bulletin*, 60, 350–362, 2010.)

industrial wastewater monitoring of PAHs. In addition to submersible fluorometers, immunoassay and molecular imprinting technology is also being incorporated with fluorescence spectroscopy for detection of PAHs [62–65].

9.4.2 Photoelectric Aerosol Sensor

PAHs readily sorb onto particles present in combustion emissions [66]. These particles, if released into the air, pose a health concern because PAHs are mutagenic and carcinogenic. As a result, the ability to detect PAHs in the air is very important. Traditionally, aerosol PAH concentrations are determined by collecting particulate matter from an airstream followed by removal of the PAHs from the particulate and quantification using GC-MS [67]. This process is laborious, expensive, and fraught with uncertainty due to the many potential sources of error.

For real-time in situ measurement of PAHs, photoelectric aerosol sensors (PASs) are used. PAH detection by PAS in air is based on the principle of photoelectric aerosol charging [68]. Figure 9.11 shows the general layout of a photoelectric aerosol sensor. First, ambient air is drawn through a quartz tube that is irradiated with ultraviolet light. The light causes PAH molecules adsorbed on carbon particulate to emit electrons that are captured by

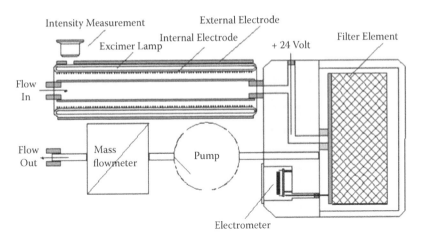

Intensity Measurement External Electrode

Excimer Lamp Internal Electrode + 24 Volt Filter Element

Flow In

Flow Out Mass flowmeter Pump

Electrometer

FIGURE 9.11
General layout of photoelectric aerosol sensor. (Image courtesy of EcoChem Analytics.)

surrounding gas molecules. The negatively charged particles are removed, and the positively charged particles are collected on a filter where an electrometer measures ion current [69]. The performance of PAS depends on particle size and molecular weight. Particles smaller than 1 μm are best for photoemission, and PAH molecules with high molecular weight (with more than four aromatic rings) promote photoionization [70].

9.4.3 Surface-Enhanced Raman Spectroscopy

Raman scattering is a vibrational spectroscopic method that can be applied for the detection of substances in both identification and quantification. This method is based on the frequency shift of the scattered radiation due to the excitation of vibrational modes on the sampled molecules [71]. Surface-enhanced Raman spectroscopy (SERS) uses nanostructured metals, such as silver and gold, to improve detection of organic molecules [72]. In particular, the discovery of the sol-gel surface-enhancing technology has allowed for the detection of low molecular weight aromatic hydrocarbons [73]. Using sol-gel technology, metal surfaces are electromagnetically enhanced to allow for the attachment of nonpolar PAHs. SERS is typically performed in laboratories and requires advanced analytical knowledge; however, attempts toward developing an in situ sensor have been researched [74, 75]. A group from the Technical University of Berlin recently deployed the first in situ SERS sensor system for the detection of PAHs in the Baltic Sea [74]. The sensor showed validity in heavily PAH-polluted waters (>150 ng (12 PAH) L^{-1}), but still needs further development for more sensitive detection.

Figure 9.12 shows the principles of SERS.

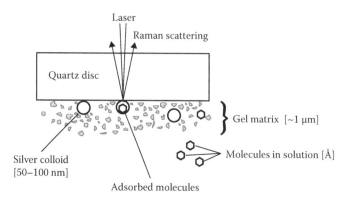

FIGURE 9.12
Principles of SERS. (From Pfannkuche et al., *Marine Pollution Bulletin*, 64, 614–626, 2012.)

9.4.4 Quartz Crystal Microbalance

A quartz crystal microbalance (QCM) utilizes the piezoelectric properties of quartz to measure the change in mass per unit area. The piezoelectric effect in quartz crystals has been known for more than a century; however, it was not used for sensing until 1959, when Sauerbray reported a linear relationship between the intrinsic crystal frequency (f:Hz) decrease of the oscillating quartz and the bound mass [76]. With the development of immunoassay and screen print technology, QCM has been used for the quantification of PAH contaminates in environmental samples [77, 78].

9.4.4.1 Immunochemical QCM

The detection of PAHs, specifically benzo[a]pyrene, using a QCM flow injection system was researched by Liu et al. [78]. They utilized the commonly used antibody mAbm10C10 for the quantification of PAH. The gold surface of the crystal was modified with a self-assembled monolayer of thiols, which was bound to benzo[a]pyrene-BSA (bovine serum albumin) using 1-ethyl-3-(3-dimethylaminopropyl)carbodichloride (EDC) and N-hydroxysuccinimide (NHS). These antibodies were displaced in the presence of the analyte benzo[a]pyrene, causing a mass change and thus a quantification of benzo[a]pyrene.

9.4.4.2 Molecularly Imprinted Polymer-Coated QCM

The detection of PAHs in the liquid phase using an organic monolayer attached to the gold electrode of a QCM was investigated by Stanley et al. [77]. They improved the selectivity to anthracene by attaching PAH acid analogues to gold-bound alkanethiol. This appended the PAH groups in a π-π stacking array, a highly attractive orientation for the anthracene analyte. Using the molecularly imprinted QCM technique, they were able to produce a sensitivity of 2 ppb for anthracene in the liquid phase.

9.4.5 Electrochemical Sensors

There are several electrochemical sensors currently in development for in situ PAH detection. One example is the electro-oxidation of PAHs in acetonitrile studied by cyclic voltammetry [79]. PAH oxidation using expanded titanium (Ti) covered with ruthenium oxide (RuO_2) electrode was also evaluated, and experimental results revealed that a current density of 9.23 mA cm^{-2} was beneficial for PAH oxidation [80]. These sensors still have challenges (e.g., representativeness, reproducibility, stability, and selectivity) to overcome prior to commercialization, but show good progress in their performance improvement and applications.

9.4.5.1 Carbon Nanotube Sensor

As a potentiometric detection of PAHs, Washe et al. [81] has shown the potential of using single-walled carbon nanotubes (SWCNTs) as sensors. The theory behind the SWCNT sensor is that π-π interactions between aromatic hydrocarbons and the CNTs will alter the chemistry of the electrical double layer and change the electrical potential [82]. This change in potential can be used to quantify aromatic hydrocarbons. A problem with this method is that it is not selective for PAH and will detect all aromatic hydrocarbons.

9.4.5.2 Lipid-Coated Mercury Sensor

Recently, a lipid-coated mercury electrode has been used to monitor PAH concentration by means of rapid cyclic voltammetry (RCV) [83]. The concept behind the sensor is that lipophilic PAH molecules will interact with the phospholipid monolayer (1,2-dioleoyl-*sn*-glycero-3-phosphochoine), changing the electrochemistry of the coating [84]. This change in chemistry will be detected by the mercury electrode and quantified by an electrosensor. This sensor is attractive because it has been shown to detect unbound PAHs; however, like the CNT sensor, other lipophilic substances (humic acid, proteins, polysaccharides) may interfere with PAH quantification [83].

9.4.5.3 Amperometric Immunosensor

An amperometric immunosensor for the detection of PAHs was developed and tested by Fähnrich et al. [85], who reported a limit of detection of 800 ppt for phenanthrene. The electrodes were created using screen printing techniques. Briefly, a silver conductor is printed on a PVC support.

Then the conductor is covered with an insulation layer, leaving two open areas. One open area is used for the electrical connection, while the other is coated with synthesized coating-conjugate (BSA-PHE: bovine serum albumin–phenanthrene) for detection of phenanthrene. The sensor showed a low limit of detection, but also showed cross-reactivates with other PAHs [85].

9.4.5.4 Capacitance Immunosensor

A capacitance immunosensor was developed by Liu et al. [86] for the detection of PAHs. The sensor used gold electrodes modified with a self-assembled monolayer (SAM) of cystamine to bind to the monoclonal antibody mAbm10C10. As the proteins (BSA-pyrene and BSA-BaP [benzo[a]pyrene]) interacted with the sensor, linear sweep voltammetry was used to detect the capacitance change. This sensor has only been tested using BSA-pyrene and BSA-BaP conjugates as analytes and still needs testing using unconjugated PAH compounds.

9.4.6 Immunoassay

Immunoassays use antibody molecules to selectively bind to specific analytes [87]. In the early 1980s, immunochemical methods were adapted for the detection of environmental contaminants. Before this, immunoassays were mainly used for clinical analysis where large biomolecules were detected. Most environmental contaminants are small molecules with molecular weights below 1000 $g \cdot mol^{-1}$. As antibodies for low molecular weight analytes developed, applications of immunoassays expanded to the field of environmental science [88]. Over a decade ago, the USEPA published standard methods for the detection of PAHs in soil (EPA Method 4035) using immunochemical methods [87]. Today, immunoassays are widely accepted for environmental applications and are commercially available. Although antibodies are typically valued for their superb specificity, antibodies to PAHs tend to recognize structurally similar PAHs, making immunoassays best for determining total PAH concentrations. In addition, immunochemical techniques often report a positive bias with PAH concentrations 10 to 100 times higher than GC and HPLC [88]. Although immunochemical methods for detecting PAHs may never replace GC and HPLC, the method has many advantages. Immunoassays are easy to use, portable, fast, and relatively inexpensive. Furthermore, many new immunochemical formats are being developed for on-site monitoring, mapping of contaminated sites, and other applications where traditional methods are not able to perform. These formats include enzyme-linked immunosorbent assay (ELISA), radioimmunoassay, piezoelectric sensing, capacitive immunosensing, surface plasmon resonance (SPR) sensing, and fluorescence sensing.

9.4.6.1 Enzyme-Linked Immunosorbent Assay

ELISA is a popular type of immunoassay used for a variety of applications, including detection of human immunodeficiency virus (HIV) antibodies in blood samples and potential food allergens in the food industry. PAH ELISA kits are commercially available for the detection of PAHs in soil, water, and sediment and have been accepted for use by the USEPA (EPA Method 4035). Traditional ELISA, developed in the early 1970s, uses chromogenic reporters and substrates that produce an observable color change to indicate the presence of an analyte; however, newer techniques that use electrochemical and fluorogenic reporters are sometimes grouped into an ELISA-type immunoassay. These new techniques are useful for real time quantification of PAHs, but are not as developed as traditional ELISA.

9.4.6.2 Other Immunoassay Formats

Although ELISA is the most commonly used immunoassay for detection of PAH, other formats are being developed. As mentioned in other sections, immunochemistry can be applied to fluoroscopic, electrochemical, surface plasmon resonance, and piezoelectric techniques for the detection of PAH compounds [63, 78, 85, 86, 88, 89].

9.4.7 Summary of PAH Sensors

Tables 9.4 to 9.6 show the summary of the developed PAH sensors, their applications, and their advantages and disadvantages.

9.5 Concluding Remarks: Outlook of In Situ PAH Monitoring Sensors

Many analytical techniques for PAH detection have been developed for decades. Among them, the development of environmental PAH sensor techniques is a revolutionary advancement in the in situ measurement. These sensors will be used to elucidate the basic mechanisms of contaminant biodegradation on the multilevel (from macro- to microscale) for contaminated soil and sediment bioremediation, as well as water body, with high spatial and temporal resolution. One of the challenges in the sensor development is to provide distinct benefits (e.g., reliable, accurate, simpler, faster, cheaper, potable, and safer) over alternative methods. The other challenge for an in situ monitoring sensor will be to

TABLE 9.4

Summary of the Developed PAH Sensors

Method	Output	PAH Sensor Type	Key Materials and Parameters	LOD	Reference
Fluoroscopy	Fluorescence intensity (a.u.)	Fluoroimmunosensor (FIS)	Anti-benzo[a]pyrene antibody	0.5 mg/L (BaP)	[62, 63]
		Submersible	Excitation wavelength 240–300 nm Emitted wavelength 310–400 nm	0.2 µg/L (phenanthrene)	[60, 61, 90]
		Molecular imprinted	Poly(vinyl chloride-co-vinyl acetate) film Excitation wavelength 337–344 nm	10 ng/L (pyrene)	[64, 65]
		CdTe quantum dots	CdTe quantum dot-modified TiO$_2$ nanotube array	3.8 ng/L (BaP)	[91]
Photoelectric aerosol sensing (PAS)	Current (mA)	PAS sensor	220 nm excimer wavelength	3 ng/M^3	[69, 92, 93]
Surface-enhanced Raman spectroscopy (SERS)	Fluorescence intensity (a.u.)	SERS sensor	Silver colloid functionalized with dimercaptoacetic calix[4]arene (DMCX) Laser diode emitting at 670.8 and 671.3 nm	20 ng/L	[89, 94]
			Gold colloids covered with per-6-deoxy-(6-thio)-β-cyclodextrin (CD-SH) Ar ion laser emitting at 488 nm	20 µg/L	[95]
			Silver colloid based on methyltriethoxysilane (MTEOS) and ethyltriethoxysilane (ETEOS) Laser emitting at 785 nm	50 ng/L	[72]

(Continued)

TABLE 9.4 *(Continued)*

Summary of the Developed PAH Sensors

Method	Output	PAH Sensor Type	Key Materials and Parameters	LOD	Reference
QCM (piezoelectric)	Frequency (Hz)	QCM immunosensor	Benzo[a]pyrene-BSA Anti-benzo[a]pyrene antibody	0.4 mg/L (BaP)	[78, 88]
		Molecular imprinted QCM	Organic monolayer attached to the surface of a gold electrode of a quartz crystal microbalance (QCM) via a covalent thiol-gold link complete with an ionically bound recognition element	2 µg/L	[65, 77, 96]
Electrochemical	Current (pA)	Amperometric immunosensor	Screen printing techniques	2 µg/L 800 ppt for phenanthrene	[85, 97]
	Capacitance (F)	Capacitive immunosensor	MAb10c10 antibody immobilized on gold electrode through a self-assembled monolayer of cystamine	1–20 mg/L	[86]
	Potential (mV)	Lipid-coated mercury	Lipid-coated mercury film electrodeposited on to platinum discs (Pt/Hg film electrode) External Ag/AgCl reference	0.1 µg/L	[83]
	Potential (mV)	Carbon nanotube	Carbon nanotubes deposited on a glassy carbon rod	3 µg/L	[81]
Surface plasmon resonance (SPR)	Intensity (a.u.)	SPR immunosensor	BaP and HBP antibodies	10 ng/L (BaP) 100 ng/L (HBP)	[98]
Other	Absorbance (abs.)	Enzyme-linked immunosorbent assay (ELISA)	Benzo[a]pyrene-BSA Anti-benzo[a]pyrene antibody	0.5 µg/L	[99]

TABLE 9.5

PAH Sensors: Applications

Method	PAH Sensor Type	Applications	Commercial Availability
Fluoroscopy	Fluoroimmunosensor (FIS)	Lab scale	N/A
	Submersible	In situ PAH detection in freshwater, ocean water, sediment, wastewater	EnviroFlu-HC (TriOS) HydroC (CONTROS) Cyclops-7 (Turner Designs)
	Molecular imprinted	Lab scale	N/A
	CdTe quantum dots	Lab scale	N/A
Photoelectric aerosol sensing (PAS)	PAS sensor	In situ PAH detection in atmosphere	PAS 2000 (EcoChem Analytics)
Surface-enhanced Raman spectroscopy (SERS)	SERS sensor	In situ PAH monitoring of ocean water and lab scale	N/A
QCM (piezoelectric)	QCM immunosensor	Lab scale	N/A
	Molecular imprinted QCM	Lab scale	N/A
Electrochemical	Amperometric immunosensor	Lab scale	N/A
	Capacitive immunosensor	Lab scale	N/A
	Lipid-coated mercury	Lab scale	N/A
SPR	SPR immunosensor	Lab scale	N/A
Other	Enzyme-linked immunosorbent assay (ELISA)	In situ PAH measuring	RaPID (Modern Water)

TABLE 9.6

PAH Sensors: Advantages and Disadvantages

Method	PAH Sensor Type	Advantages	Disadvantages
Fluoroscopy	Fluoroimmunosensor (FIS)	High sensitivity Simple measurement	Antibody cross-reactivity Short shelf life
	Submersible	Quick response time Portable and commercially available Can be applied to multiple environments	Interference by other aromatic molecules (e.g., humic acid) Inaccurate below 1 ppb
	Molecular imprinted	High sensitivity Low humic acid interference	Complicated manufacturing process
	CdTe quantum dots	Enhanced fluorescent intensity High sensitivity Little interference	Potential pollution from the device
Photoelectric aerosol sensing (PAS)	PAS sensor	Inexpensive Total PAH detection In situ atmospheric PAH measurement	Only measures particle-bound PAH Selective for particle sizes less than 1 μm
Surface-enhanced Raman spectroscopy (SERS)	SERS sensor	In situ PAH measurement applied in seawater High durability Measures many types of PAHs SERS is a developed analytical method	In situ device is relatively new and needs further development Large and costly Requires advanced analytical knowledge
QCM (piezoelectric)	QCM immunosensor	Provides direct study of antibody interactions Direct mass measurement	Low selectivity Low flexibility Only analyzed in lab
	Molecular imprinted QCM	Direct mass measurement	Low yield Low sensitivity in mass detection Low selectivity Only analyzed in lab
Electrochemical	Amperometric immunosensor	Small portable design	Long sample preparation time Short life time Low selectivity
	Capacitive immunosensor	Simple measurement	Long response time Low selectivity
	Lipid-coated mercury	Low detection limit	Long detection time Interference from organic matter exists

(Continued)

TABLE 9.6 *(Continued)*

PAH Sensors: Advantages and Disadvantages

Method	PAH Sensor Type	Advantages	Disadvantages
SPR	SPR immunosensor	High sensitivity No interference or cross-activity	Long response time
Other	Enzyme-linked immunosorbent assay (ELISA)	Overall PAH measurement Commercially available EPA approved method	Consumes reagents

develop efficient surface modification of sensing materials for antifouling (e.g., functionalized and controlled surface design or biomimetics) to prevent biofouling during in situ monitoring and provide more reliable sensor surface technology.

References

1. S.K. Samanta, O.V. Singh, and R.K. Jain, Polycyclic aromatic hydrocarbons: Environmental pollution and bioremediation, *TRENDS in Biotechnology*, 20, 243–248, 2002.
2. R.G. Harvey, Mechanisms of carcinogenesis of polycyclic aromatic hydrocarbons, *Polycyclic Aromatic Compounds*, 9, 1–23, 1996.
3. U.S. Environmental Protection Agency, Drinking water standards, http://www.epa.gov/safewater/standards.html (accessed June 2014).
4. European Communities, Council Directive 98/83/EC of 3 November 1998 on the quality of water intended for human consumption, *Official Journal of the European Communities*, L-330 (December 5, 1998).
5. World Health Organization, Guidelines for drinking water quality: Health criteria and other supporting information, 2nd ed., Geneva, Switzerland, 1996.
6. M.D. Watson, A. Fechtenkötter, and K. Müllen, Big is beautiful—Aromaticity revisited from the viewpoint of macromolecular and supramolecular benzene chemistry, *Chemical Reviews*, 101, 1267–1300, 2001.
7. J. Juselius and D. Sundholm, Polycyclic antiaromatic hydrocarbons, *Physical Chemistry Chemical Physics*, 10, 6630–6634, 2008.
8. T. Krygowski, J. Zachara, B. Osmialowski, and R. Gawinecki, Topology-driven physicochemical properties of π-electron systems. 1. Does the Clar rule work in cyclic π-electron systems with the intramolecular hydrogen or lithium bond? *Journal of Organic Chemistry*, 71, 7678–7682, 2006.
9. Y. Ruiz-Morales, The agreement between Clar structures and nucleus-independent chemical shift values in pericondensed benzenoid polycyclic aromatic hydrocarbons: An application of the Y-rule, *Journal of Physical Chemistry A*, 108, 10873–10896, 2004.

10. J.M. Schulman and R.L. Disch, Aromatic character of [n]helicenes and [n]phenacenes, *Journal of Physical Chemistry A*, 103, 6669–6672, 1999.

11. R. Dabestani and I.N. Ivanov, A compilation of physical, spectroscopic and photophysical properties of polycyclic aromatic hydrocarbons, *Photochemistry and Photobiology*, 70, 10–34, 1999.

12. National Climatic Data Center (NCDC), National Oceanic and Atmospheric Administration (NOAA), U.S. Department of Commerce, http://www.ncdc.noaa.gov/.

13. E.S. Pysh and N.C. Yang, Polarographic oxidation potentials of aromatic compounds, *Journal of the American Chemical Society*, 85, 2124–2130, 1963.

14. A. Zweig and J.E. Lehnsen, Cumulative influence of methylthio groups on the π-system properties of aromatic hydrocarbons, *Journal of the American Chemical Society*, 87, 2647–2657, 1965.

15. A. Zweig, A.H. Maurer, and B.G. Roberts, Oxidation reduction and electrochemiluminescence of donor-substituted polycyclic aromatic hydrocarbons, *Journal of Organic Chemistry*, 32, 1322–1329, 1967.

16. C.R. Johnson and S.A. Asher, A new selective technique for characterization of polycyclic aromatic-hydrocarbons in complex samples—UV resonance Raman spectrometry of coal liquids, *Analytical Chemistry*, 56, 2258–2261, 1984.

17. R.N. Okparanma and A.M. Mouazen, Determination of total petroleum hydrocarbon (TPH) and polycyclic aromatic hydrocarbon (PAH) in soils: A review of spectroscopic and nonspectroscopic techniques, *Applied Spectroscopy Reviews*, 48, 458–486, 2013.

18. R. Schoental and E.J.Y. Scott, Fluorescence spectra of polycyclic aromatic hydrocarbons in solution, *Journal of the Chemical Society*, 1683–1696, 1949.

19. S.A. Tucker, W.E. Acree, M.J. Tanga, S. Tokita, K. Hiruta, and H. Langhals, Spectroscopic properties of polycyclic aromatic compounds: Examination of nitromethane as a selective fluorescence quenching agent for alternant polycyclic aromatic nitrogen hetero-atom derivatives, *Applied Spectroscopy*, 46, 229–235, 1992.

20. T.L. Cecil and S.C. Rutan, Correction for fluorescence response shifts in polyaromatic hydrocarbon mixtures with an innovations-based Kalman filter method, *Analytical Chemistry*, 62, 1998–2004, 1990.

21. J.R. Lakowicz, *Principles of fluorescence spectroscopy*, 3rd ed., Springer, New York, 2006.

22. S.A. Tucker, W.E. Acree, B.P. Cho, R.G. Harvey, and J.C. Fetzer, Spectroscopic properties of polycyclic aromatic-hydrocarbons—Effect of solvent polarity on the fluorescence emission behavior of select fluoranthene, fluorenochrysene, indenochrysene, and indenopyrene derivatives, *Applied Spectroscopy*, 45, 1699–1705, 1991.

23. M.E. Sigman, J.T. Barbas, E.A. Chevis, and R. Dabestani, Spectroscopy and photochemistry of 1-methoxynaphthalene on SiO2, *New Journal of Chemistry*, 20, 243–248, 1996.

24. J.T. Barbas, M.E. Sigman, and R. Dabestani, Photochemical oxidation of phenanthrene sorbed on silica gel, *Environmental Science and Technology*, 30, 1776–1780, 1996.

25. J.T. Barbas, M.E. Sigman, R. Arce, and R. Dabestani, Spectroscopy and photochemistry of fluorene at a silica gel/air interface, *Journal of Photochemistry and Photobiology A*, 109, 229–236, 1997.

26. M.E. Sigman, S. Read, J.T. Barbas, I. Ivanov, E.W. Hagmann, A.C. Buchanan, et al., Rapid molecular motion of pyrene and benzene moieties covalently attached to silica surfaces, *Journal of Physical Chemistry A*, 107, 3450–3456, 2003.

27. R. Dabestani, K.J. Ellis, and M.E. Sigman, Photodecomposition of Anthracene on dry surfaces—Products and mechanism, *Journal of Photochemistry and Photobiology A*, 86, 231–239, 1995.

28. J.T. Barbas, R. Dabestani, and M.E. Sigman, A mechanistic study of photode-composition of acenaphthylene on a dry silica surface, *Journal of Photochemistry and Photobiology A*, 80, 103–111, 1994.

29. C. Reyes, M.E. Sigman, R. Arce, J.T. Barbas, and R. Dabestani, Photochemistry of acenaphthene at a silica gel/air interface, *Journal of Photochemistry and Photobiology A*, 112, 277–283, 1998.

30. R. Dabestani and I.N. Ivanov, A compilation of physical, spectroscopic and pho-tophysical properties of polycyclic aromatic hydrocarbons, *Photochemistry and Photobiology*, 70, 10–34, 1999.

31. C.H. Lochmuller and T.J. Wenzel, Spectroscopic studies of pyrene at silica interfaces, *Journal of Physical Chemistry*, 94, 4230–4235, 1990.

32. U.S. Environmental Protection Agency, EPA Office of Research and Development, Handbook—Ground water: Methodology, Vol. II, EPA 625/6-90/016b, 1991.

33. U.S. Environmental Protection Agency, EPA Office of Research and Development, Methods for the determination of organic compounds in drinking water, EPA 600/ 4–88/039, 1991.

34. L. Oliferova, M. Statkus, G. Tsysin, O. Shpigun, and Y. Zolotov, On-line solid-phase extraction and HPLC determination of polycyclic aromatic hydrocarbons in water using fluorocarbon polymer sorbents, *Analytica Chimica Acta*, 538, 35–40, 2005.

35. F. Busetti, A. Heitz, M. Cuomo, S. Badoer, and P. Traverso, Determination of sixteen polycyclic aromatic hydrocarbons in aqueous and solid samples from an Italian wastewater treatment plant, *Journal of Chromatography A*, 1102, 104–115, 2006.

36. W.D. Wang, Y.M. Huang, W.Q. Shu, and J. Cao, Multiwalled carbon nanotubes as adsorbents of solid-phase extraction for determination of polycyclic aromatic hydrocarbons in environmental waters coupled with high-performance liquid chromatography, *Journal of Chromatography A*, 1173, 27–36, 2007.

37. H. Wang and A.D. Campiglia, Determination of polycyclic aromatic hydro-carbons in drinking water samples by solid-phase nanoextraction and high-performance liquid chromatography, *Analytical Chemistry*, 80, 8202–8209, 2008.

38. C. Gooijer, F. Ariese, and J.W. Hofstraat, eds., *Shpol'skii spectroscopy and other site-selection methods: Applications in environmental analysis, bioanalytical chemistry, and chemical physics*, Wiley–Interscience, New York, 2000.

39. A.J. Bystol, A.D. Campiglia, and G.D. Gillispie, Time-resolved laser-excited Shpol'skii spectrometry with a fiber-optic probe and ICCd camera, *Applied Spectroscopy*, 54, 910–917, 2000.

40. A.J. Bystol, A.D. Campiglia, and G.D. Gillispie, Laser-induced multidimen-sional fluorescence spectroscopy in Shpol'skii matrixes with a fiber optic probe at liquid helium temperature, *Analytical Chemistry*, 73, 5762–5770, 2001.

41. A.F. Moore, F. Barbosa, and A.D. Campiglia, Combining cryogenic fiber optic probes with commercial spectrofluorimeters for the synchronous fluorescence Shpol'skii spectroscopy of high molecular weight polycyclic aromatic hydrocar-bons, *Applied Spectroscopy*, 68, 14–25, 2014.

42. A.D. Campiglia, A.J. Bystol, and S.J. Yu, Instrumentation for multidimensional luminescence spectroscopy and its application to low-temperature analysis in Shpol'skii matrixes and optically scattering media, *Analytical Chemistry*, 78, 484–492, 2006.

43. A.J. Bystol, T. Thorstenson, and A.D. Campiglia, Laser-induced multidimensional fluorescence spectroscopy in Shpol'skii matrixes for the analysis of polycyclic aromatic hydrocarbons in HPLC fractions and complex environmental extracts, *Environmental Science and Technology*, 36, 4424–4429, 2002.

44. A.J. Bystol, S. Yu, and A.D. Campiglia, Analysis of polycyclic aromatic hydrocarbons in HPLC fractions by laser-excited time-resolved Shpol'skii spectrometry with cryogenic fiber-optic probes, *Talanta*, 60, 449–458, 2003.

45. S. Yu and A.D. Campiglia, Laser-excited time-resolved Shpol'skii spectroscopy for the direct analysis of dibenzopyrene isomers in liquid chromatography fractions, *Applied Spectroscopy*, 58, 1385–1393, 2004.

46. W.B. Wilson and A.D. Campiglia, Analysis of co-eluted isomers of high-molecular weight polycyclic aromatic hydrocarbons in high performance liquid chromatography fractions via solid-phase nanoextraction and time-resolved Shpol'skii spectroscopy, *Journal of Chromatography A*, 1218, 6922–6929, 2011.

47. A.J. Bystol, J.L. Whitcomb, and A.D. Campiglia, Solid-liquid extraction laser excited time-resolved Shpol'skii spectrometry: A facile method for the direct identification of fifteen priority pollutants in water samples, *Environmental Science and Technology*, 35, 2566–2571, 2001.

48. S. Yu and A.D. Campiglia, Direct determination of dibenzo[a,l]pyrene and its four dibenzopyrene isomers in water samples by solid-liquid extraction and laser-excited time-resolved Shpol'skii spectrometry, *Analytical Chemistry*, 77, 1440–1447, 2005.

49. H. Wang, S. Yu, and A.D. Campiglia, Solid-phase nano-extraction and laser-excited time-resolved Shpol'skii spectroscopy for the analysis of polycyclic aromatic hydrocarbons in drinking water samples, *Analytical Biochemistry*, 385, 249–256, 2009.

50. H. Wang and A.D. Campiglia, Direct determination of benzo[a]pyrene in water samples by a gold nanoparticle-based solid-phase extraction method and laser-excited time-resolved Shpol'skii spectrometry, *Talanta*, 83, 233–240, 2010.

51. W.B. Wilson and A.D. Campiglia, Determination of polycyclic aromatic hydrocarbons with molecular weight 302 in water samples by solid-phase nano-extraction and laser excited time-resolved Shpol'skii spectroscopy, *The Analyst*, 136, 3366–3374, 2011.

52. Y. Wang, Z. Wang, M. Ma, C. Wang, and Z. Mo, Monitoring priority pollutants in a sewage treatment process by dichloromethane extraction and triolein-semipermeable membrane device (SPMD), *Chemosphere*, 43, 339–346, 2001.

53. I. Urbe and J. Ruana, Application of solid-phase extraction discs with a glass fiber matrix to fast determination of polycyclic aromatic hydrocarbons in water, *Journal of Chromatography A*, 778, 337–345, 1997.

54. S. Hawthorne, S. Trembley, C. Moniot, C. Grabanski, and D. Miller, Static subcritical water extraction with simultaneous solid-phase extraction for determining polycyclic aromatic hydrocarbons on environmental solids, *Journal of Chromatography A*, 886, 237–244, 2000.

55. A. King, J. Readman, and J. Zhou, Determination of polycyclic aromatic hydrocarbons in water by solid-phase microextraction–gas chromatography–mass spectrometry, *Analytica Chimica Acta*, 523, 259–267, 2004.

56. J. Chen and J.B. Pawliszyn, Solid phase microextraction coupled to high-performance liquid chromatography, *Analytical Chemistry*, 67, 2530–2533, 1995.

57. M.D. Burford, S.B. Hawthorne, and D.J. Miller, Extraction rates of spiked versus native PAHs from heterogeneous environmental samples using supercritical fluid extraction and sonication in methylene chloride, *Analytical Chemistry*, 65, 1497–1505, 1993.

58. H. Bagheri and A. Mohammadi, Pyrrole-based conductive polymer as the solid-phase extraction medium for the preconcentration of environmental pollutants in water samples followed by gas chromatography with flame ionization and mass spectrometry detection, *Journal of Chromatography A*, 1015, 23–30, 2003.

59. N. Hudson, A. Baker, and D. Reynolds, Fluorescence analysis of dissolved organic matter in natural, waste and polluted waters—A review, *River Research and Applications*, 23, 631–649, 2007.

60. M. Tedetti, C. Guigue, and M. Goutx, Utilization of a submersible UV fluorometer for monitoring anthropogenic inputs in the Mediterranean coastal waters, *Marine Pollution Bulletin*, 60, 350–362, 2010.

61. M. Tedetti, P. Joffre, and M. Goutx, Development of a field-portable fluorometer based on deep ultraviolet LEDs for the detection of phenanthrene- and tryptophan-like compounds in natural waters, *Sensors and Actuators B: Chemical*, 182, 416–423, 2013.

62. T. Vo-Dinh, B. Tromberg, G. Griffin, K. Ambrose, M. Sepaniak, and E. Gardenhire, Antibody-based fiberoptics biosensor for the carcinogen benzo(a)pyrene, *Applied Spectroscopy*, 41, 735–738, 1987.

63. A. Ius, M. Bacigalupo, A. Roda, and C. Vaccari, 006 development of a time-resolved fluoroimmunoassay of benzo(a)pyrene in water, *Fresenius' Journal of Analytical Chemistry*, 343, 55–56, 1992.

64. C. Sluszny, V.V. Gridin, V. Bulatov, and I. Schechter, Polymer film sensor for sampling and remote analysis of polycyclic aromatic hydrocarbons in clear and turbid aqueous environments, *Analytica Chimica Acta*, 522, 145–152, 2004.

65. F.L. Dickert, P. Achatz, and K. Halikias, Double molecular imprinting—A new sensor concept for improving selectivity in the detection of polycyclic aromatic hydrocarbons (PAHs) in water, *Fresenius' Journal of Analytical Chemistry*, 371, 11–15, 2001.

66. J.C. Dunbar, C.-I. Lin, I. Vergucht, J. Wong, and J.L. Durant, Estimating the contributions of mobile sources of PAH to urban air using real-time PAH monitoring, *Science of the Total Environment*, 279, 1–19, 2001.

67. G. Agnesod, R. De Maria, M. Fontana, and M. Zublena, Determination of PAH in airborne particulate: Comparison between off-line sampling techniques and an automatic analyser based on a photoelectric aerosol sensor, *Science of the Total Environment*, 189, 443, 1996.

68. H. Burtscher, Measurement and characteristics of combustion aerosols with special consideration of photoelectric charging and charging by flame ions, *Journal of Aerosol Science*, 23, 549–595, 1992.

69. K. Hart, S.R. McDow, W. Giger, D. Steiner, and H. Burtscher, The correlation between in-situ, real-time aerosol photoemission intensity and particulate polycyclic aromatic hydrocarbon concentration in combustion aerosols, *Water, Air, and Soil Pollution*, 68, 75–90, 1993.

70. S.R. McDow, W. Giger, H. Burtscher, A. Schmidt-Ott, and H.C. Siegmann, Polycyclic aromatic hydrocarbons and combustion aerosol photoemission, *Atmospheric Environment A*, 24, 2911–2916, 1990.

71. M. Fleischmann, P. Hendra, and A. McQuillan, Raman spectra of pyridine adsorbed at a silver electrode, *Chemical Physics Letters*, 26, 163–166, 1974.

72. H. Schmidt, N. Bich Ha, J. Pfannkuche, H. Amann, H.-D. Kronfeldt, and G. Kowalewska, Detection of PAHs in seawater using surface-enhanced Raman scattering (SERS), *Marine Pollution Bulletin*, 49, 229–234, 2004.

73. T. Murphy, H. Schmidt, and H.-D. Kronfeldt, Use of sol-gel techniques in the development of surface-enhanced Raman scattering (SERS) substrates suitable for in situ detection of chemicals in sea-water, *Applied Physics B*, 69, 147–150, 1999.

74. J. Pfannkuche, L. Lubecki, H. Schmidt, G. Kowalewska, and H.-D. Kronfeldt, The use of surface-enhanced Raman scattering (SERS) for detection of PAHs in the Gulf of Gdańsk (Baltic Sea), *Marine Pollution Bulletin*, 64, 614–626, 2012.

75. O. Péron, E. Rinnert, F. Colas, M. Lehaitre, and C. Compère, First steps of in situ surface-enhanced Raman scattering during shipboard experiments, *Applied Spectroscopy*, 64, 1086–1093, 2010.

76. K.A. Marx, Quartz crystal microbalance: A useful tool for studying thin polymer films and complex biomolecular systems at the solution-surface interface, *Biomacromolecules*, 4, 1099–1120, 2003.

77. S. Stanley, C. Percival, M. Auer, A. Braithwaite, M. Newton, G. McHale, and W. Hayes, Detection of polycyclic aromatic hydrocarbons using quartz crystal microbalances, *Analytical Chemistry*, 75, 1573–1577, 2003.

78. M. Liu, Q.X. Li, and G.A. Rechnitz, Flow injection immunosensing of polycyclic aromatic hydrocarbons with a quartz crystal microbalance, *Analytica Chimica Acta*, 387, 29–38, 1999.

79. M.E. Peover and B.S. White, The electro-oxidation of polycyclic aromatic hydrocarbons in acetonitrile studied by cyclic voltammetry, *Journal of Electroanalytical Chemistry and Interfacial Electrochemistry*, 13(1), 93–99, 1967.

80. L.-H. Tran, P. Drogui, G. Mercier, and J.-F. Blais, Electrochemical degradation of polycyclic aromatic hydrocarbons in creosote solution using ruthenium oxide on titanium expanded mesh anode, *Journal of Hazardous Materials*, 164(2), 1118–1129, 2009.

81. A.P. Washe, S. Macho, G.A. Crespo, and F.X. Rius, Potentiometric online detection of aromatic hydrocarbons in aqueous phase using carbon nanotube-based sensors, *Analytical Chemistry*, 82, 8106–8112, 2010.

82. A. Star, J.-C.P. Gabriel, K. Bradley, and G. Grüner, Electronic detection of specific protein binding using nanotube FET devices, *Nano Letters*, 3, 459–463, 2003.

83. A. Penezic, B. Gasparovic, D. Stipanicev, and A. Nelson, In-situ electrochemical method for detecting freely dissolved polycyclic aromatic hydrocarbons in waters, *Environmental Chemistry*, 11, 173–180, 2013.

84. A. Nelson and F. Leermakers, Substrate-induced structural changes in electrode-adsorbed lipid layers: Experimental evidence from the behaviour of phospholipid layers on the mercury-water interface, *Journal of Electroanalytical Chemistry and Interfacial Electrochemistry*, 278, 73–83, 1990.

85. K. Fähnrich, M. Pravda, and G. Guilbault, Disposable amperometric immunosensor for the detection of polycyclic aromatic hydrocarbons (PAHs) using screen-printed electrodes, *Biosensors and Bioelectronics*, 18, 73–82, 2003.
86. M. Liu, G.A. Rechnitz, K. Li, and Q.X. Li, Capacitive immunosensing of polycyclic aromatic hydrocarbon and protein conjugates, *Analytical Letters*, 31, 2025–2038, 1998.
87. D. Barceló, A. Oubina, J. Salau, and S. Perez, Determination of PAHs in river water samples by ELISA, *Analytica Chimica Acta*, 376, 49–53, 1998.
88. K.A. Fähnrich, M. Pravda, and G.G. Guilbault, Immunochemical detection of polycyclic aromatic hydrocarbons (PAHs), *Analytical Letters*, 35, 1269–1300, 2002.
89. Y.-H. Kwon, A. Kolomijeca, K. Sowoidnich, and H.-D. Kronfeldt, High sensitivity calixarene SERS substrates for the continuous in-situ detection of PAHs in seawater, in *SPIE defense, security, and sensing*, International Society for Optics and Photonics, 2011, pp. 80240E–80240E-9.
90. M.L. Nahorniak and K.S. Booksh, Excitation-emission matrix fluorescence spectroscopy in conjunction with multiway analysis for PAH detection in complex matrices, *Analyst*, 131, 1308–1315, 2006.
91. L. Yang, B. Chen, S. Luo, J. Li, R. Liu, and Q. Cai, Sensitive detection of polycyclic aromatic hydrocarbons using CdTe quantum dot-modified TiO2 nanotube array through fluorescence resonance energy transfer, *Environmental Science and Technology*, 44, 7884–7889, 2010.
92. E. Velasco, P. Siegmann, and H.C. Siegmann, Exploratory study of particle-bound polycyclic aromatic hydrocarbons in different environments of Mexico City, *Atmospheric Environment*, 38, 4957–4968, 2004.
93. J.W. Childers, C.L. Witherspoon, L.B. Smith, and J.D. Pleil, Real-time and integrated measurement of potential human exposure to particle-bound polycyclic aromatic hydrocarbons (PAHs) from aircraft exhaust, *Environmental Health Perspectives*, 108, 853, 2000.
94. L. Guerrini, J.V. Garcia-Ramos, C. Domingo, and S. Sanchez-Cortes, Sensing polycyclic aromatic hydrocarbons with dithiocarbamate-functionalized Ag nanoparticles by surface-enhanced Raman scattering, *Analytical Chemistry*, 81, 953–960, 2009.
95. Y. Xie, X. Wang, X. Han, W. Song, W. Ruan, J. Liu, B. Zhao, and Y. Ozaki, Selective SERS detection of each polycyclic aromatic hydrocarbon (PAH) in a mixture of five kinds of PAHs, *Journal of Raman Spectroscopy*, 42, 945–950, 2011.
96. F.L. Dickert, H. Besenböck, and M. Tortschanoff, Molecular imprinting through van der Waals interactions: Fluorescence detection of PAHs in water, *Advanced Materials*, 10, 149–151, 1998.
97. A. Ahmad and E. Moore, Electrochemical immunosensor modified with self-assembled monolayer of 11-mercaptoundecanoic acid on gold electrodes for detection of benzo [a] pyrene in water, *Analyst*, 137, 5839–5844, 2012.
98. K.V. Gobi and N. Miura, Highly sensitive and interference-free simultaneous detection of two polycyclic aromatic hydrocarbons at parts-per-trillion levels using a surface plasmon resonance immunosensor, *Sensors and Actuators B: Chemical*, 103, 265–271, 2004.
99. A. Székács, H. Le, D. Knopp, and R. Niessner, A modified enzyme-linked immunosorbent assay (ELISA) for polyaromatic hydrocarbons, *Analytica Chimica Acta*, 399, 127–134, 1999.

10

PAHs in Sewage Sludge, Soils, and Sediments

Amy J. Forsgren

Xylem Inc., Sundbyberg, Sweden

CONTENTS

This chapter discusses polycyclic aromatic hydrocarbons (PAHs) in sewage sludge, and the related topics of PAHs in soils and sediments. Soils are often amended with sewage sludge because it contains nutrients. Sediments—in rivers, lakes, harbors, and estuaries—tend to accumulate the PAHs that exited the wastewater treatment plant (WWTP) in the liquid effluent. Soils that have been amended with sludge can end up in water resources due to soil erosion. Sediments can also become soils, when sediment dredged from harbors or rivers is deposited on land.

Sludge, soils, and sediments have some similarity in their ability to use bioremediation to reduce organic contaminants.

10.1 Amounts of PAHs in Sludge

Sludge is the residue collected at the primary (physical/chemical), secondary (biological), and tertiary stages of the WWTP process. Sludge is made up of organic and inorganic solids, WWTP process additives, biomass formed during the aerobic or anaerobic degradation processes at the WWTP, and water (Schowanek et al. 2004). It contains some compounds of high agricultural value, such as phosphorus, nitrogen, potassium, and organic matter, as well as smaller amounts of calcium, sulfur, and magnesium. It also tends to contain undesirable elements, such as heavy metals, pathogens, and organic pollutants.

The character or quality of the sludge depends foremost on the pollution load of the water being treated; other important factors are the wastewater treatment process and the sludge treatment process. Table 10.1 briefly summarizes the amounts of PAHs reported in sewage sludge for several studies.

10.1.1 Why So Many Sludge Studies?

There is no typical sludge treatment layout; configurations at WWTPs depend on the capacity of the WWTP, the expected quality of the water to be treated, and the emission and disposal regulations for the particular WWTP or locality (Mailler et al. 2014).

This helps explain why there are a number of studies characterizing sludge. When every WWTP can yield different data in analyses, a study of each WWTP becomes relevant and important to the process owners and the authorities safeguarding local natural resources.

Since there are an estimated 16,000 WWTPs in the United States alone, the study of sludge quality, and particularly PAHs, in the technical literature is, if anything, underrepresented.

10.2 Sludge Treatment

Sludge is usually treated before recycling or final disposal for a number of reasons:

To render pathogens harmless

To reduce its water content

To reduce the sludge's propensity toward fermentation

There are several sludge treatment processes that can be combined or used singly: thickening, dewatering, stabilization, disinfection, and thermal drying. These are described in Table 10.2.

TABLE 10.1

Amounts of PAHs Reported in Sewage Sludge

Location	Total PAHs in Sludge, μg/kg Dried Weight	Notes	Reference
San Diego, California, 1 WWTP	5920–17,500		Zeng and Vista 1997
Cluj Napoca, Romania, 1 WWTP	15.61	Major component: napthalene (11.50)	Alhafez et al. 2013
Paris, France, 1 WWTP	14,000–31,000	Seine Aval WWTP receives estimated 26.7 kg/year of PAHs (draining an industrial area)	Blanchard et al. 2004
Lombardy, Italy, 1 WWTP	2405 (secondary sludge); 2645 (final sludge)	Major component: pyrene	Torretta and Katsoyiannis 2013
Venice, Italy, 1 WWTP	1260–1440	Major components: pyrene, benzo(a)anthracene, chrysene	Busetti et al. 2006
West Poland, North Poland, 2 WWTPs	2433	WWTP with mechanical and biological system	Werle and Dudziak 2014
West Poland, North Poland, 2 WWTPs	621	WWTP with mechanical, biological, and chemical system with simultaneous phosphorus precipitation	Werle and Dudziak 2014
Southeast Poland, 5 WWTPs	3674–11,236		Oleszczuk 2007
Cádiz, Spain, 1 WWTP	1945 (March); 10,100 (June)	Primary sludge	Villar et al. 2006
Spain, 6 WWTPs	1130–5520	Could not see a relation between sludge treatment process used and PAH levels	Pérez et al. 2001
Norway, 9 WWTPs	700–30,000	Mean = 3900	Augulyte 2001
Umeå, Sweden	410–2500		NVV 2003
Västra Götaland County, Sweden, 19 WWTPs	1100		
Roskilde, Denmark, 1 WWTP	4132	Measured PAHs + nitro-PAHs (NPAHs); 1021 were N-PAHs; 3111 were PAHs	Vikelsoe et al. 2002

(Continued)

TABLE 10.1 (*Continued*)

Amounts of PAHs Reported in Sewage Sludge

Location	Total PAHs in Sludge, μg/kg Dried Weight	Notes	Reference
Mainland China and Hong Kong, 11 WWTPs	1400–33,000	Highest values found in Beijing	Cai et al. 2007
Beijing, China, 6 WWTPs	2467–25,923	Individual PAH contents varied among the 6 plants, but ratios indicate petroleum and fossil fuel combustion are dominant contributors	Dai et al. 2007
Guangdong Province, China, 19 WWTPs in 6 cities	177.2–4421.8	Major components: phenanthrene, fluoranthene, pyrene, chrysene	Zeng et al. 2012
South Korea, 4 WWTPs	340–3850		Ju et al. 2009
Tunisia, 9 WWTPs	96–7718	Highest PAH content from natural lagooning (untreated WW)	Khadhar et al. 2010

TABLE 10.2

Examples of Sludge Treatment Processes

Process	Description
Sludge pasteurization	Minimum of 20 min at 70°C or minimum of 4 h at 55°C, followed in all cases by primary mesophilic anaerobic digestion.
Mesophilic anaerobic digestion	Mean retention period of at least 12 or 24 days primary digestion in temperature range 35 ± 3 or 25 ± 3°C, respectively, followed by a stage providing a mean retention period of at least 14 days.
Thermophilic aerobic digestion	Mean retention period of at least 7 days digestion. All sludge to be subject to a minimum of 55°C for a period of maturation adequate to ensure that compost reaction process is substantially complete.
Composting (windows and aerated piles)	The compost must be maintained at 40°C at least 5 days and for 4 h during this period at a minimum of 55°C within the body of the pile, followed by a period of maturation adequate to ensure that the compost reaction process is substantially complete.
Lime stabilization of liquid sludge	Addition of lime to raise pH to greater than 12.0 and sufficient to ensure that the pH is not less than 12 for a minimum period of 2 h. The sludge can then be used directly.
Liquid storage	Storage of retreated liquid sludge for a minimum period of 3 months.
Dewatering and storage	Conditioning of untreated sludge with lime, followed by dewatering and storage of the cake for a minimum period of 3 months. Storage for a period of 14 days as sludge has been subject to primary mesophilic anaerobic digestion.

Source: Fytili and Zabaniotou, *Renewable and Sustainable Energy Reviews*, 12(1), 116–140, 2008.

Mailler et al. (2014) studied sludge treatment at three WWTPs in the Paris area. Because they all receive water from the Paris catchment, the incoming wastewater is believed to be roughly similar. The WWTPs have different sludge treatment processes, which led to different PAH levels in the final sludge. Thermal drying seems to allow a decrease in PAHs in the final sludge, but probably by volatizing the compounds—moving from one sort of emission to another. Overall, digestion was the only treatment that led to a real reduction in micropollutant loads. The sludge treatment processes of these three Parisian WWTPs are shown in Figure 10.1.

10.3 Sludge Disposal

The handling of sewage sludge is one of the most significant challenges to the wastewater treatment industry. Fytili and Zabaniotou (2008) summed it up well: "Without a reliable disposal method for the sludge, the actual concept of water protection will fail."

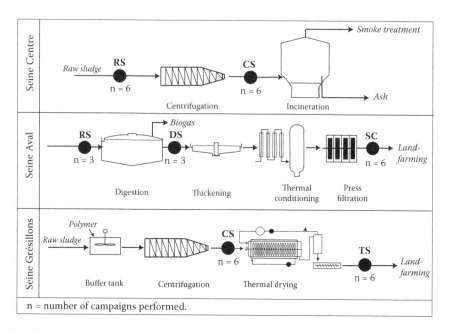

FIGURE 10.1
Sludge treatment processes in three Paris WWTPs. (Reprinted with permission from Mailler, R., et al., *Waste Management*, 34(7), 1271–1226.)

How much sludge are we talking about? In Europe it has been estimated that 90 g dw of sewage sludge is produced per person per day, in the primary, secondary, and tertiary treatments of WWTPs (Davis 1996). In Japan in 2004, approximately 244,000,000 m^3 of sewage sludge was produced (Hong et al. 2009).

10.3.1 Disposal Methods

The primary methods for disposal of sewage sludge are:

Incineration

Recycling to agriculture, also called soil amendment or land spreading

Landfill/sanitary landfill

Ocean dumping (no longer an option in many parts of the world)

How much of each disposal method is used depends on the location and especially legislations in effect. Ocean dumping will not be covered in this chapter; it was outlawed in the United States in 1992 and in the EU in 1998. Figure 10.2 shows the percentage of sludge in the EU that is disposed of by incineration, landfill, agriculture, and sylviculture/land reclamation.

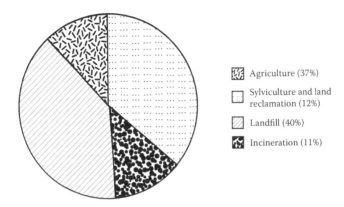

Agriculture (37%)

Sylviculture and land
reclamation (12%)

Landfill (40%)

Incineration (11%)

FIGURE 10.2
Sludge disposal within the EU. (Data from Fytili and Zabaniotou, *Renewable and Sustainable Energy Reviews*, 12(1), 116–140, 2008.)

10.3.2 Incineration

Incineration is the term encompassing several thermal processing technologies (Werther and Ogada 1999; EC 2001; Fytili and Zabaniotou 2008):

Monocombustion: Incinerators include multiple hearth, fluidized bed, smelting furnace.

Co-combustion: Examples include (1) co-combustion with coal in fluidized bed or pulverized bed coal combustors and (2) co-incineration with municipal waste in furnaces.

Alternative processes: The driver for developing these processes tends to be the cost of treating flue gas and ash in combustion processes. Alternative processes include wet oxidation, pyrolysis, gasification, and melting.

In wet oxidation, sludge in the aqueous phase is subjected to high temperatures (150 to 330°C), very high pressure (1 to 22 MPa), and an oxygen-rich environment. The organic content of the sludge is thermally degraded, hydrolyzed, and oxidized.

In pyrolysis, sludge is heated in an inert atmosphere at 300 to 900°C. In the oxygen-free atmosphere, organic content undergoes thermal cracking reactions and condensation reactions. Products are:

Gaseous: Methane, hydrogen, carbon dioxide, carbon monoxide, and small amounts of other gases.

Liquid: Tar or oil or both; acetic acid, acetone, and methanol are frequently present.

Solid: Char, almost entirely carbon with small amounts of inert materials.

In gasification, the organic content of sludge is converted to combustible gas and ash in a reducing atmosphere. Because it is done in an oxygen-poor atmosphere, NOx and SOx emissions are limited compared to conventional combustion.

Dewatered sludge melting consists of burning the sludge at 1300 to 1800°C, instead of the 800 to 900°C of conventional combustion. The increase in temperature leads to higher running costs. The advantages gained are that the solid product is a slag that is expected to have much less leaching when landfilled than conventional ash, metals recycling is a possibility, and the amounts of solids produced are small (Hong et al. 2009).

Combustion, pyrolysis, and wet oxidation remove the organic content of the sludge and leave only ash for disposal. The product of gasification, synthesis gas, can be used either as fuel or as a raw material for chemical industries (Nowicki et al. 2011).

Incineration is often the most attractive disposal method, as legal limitations make landfill and agricultural reuse more difficult (and since sea disposal is no longer an option). In Japan incineration accounts for 55% of sludge disposal; in the United States it is 25%, and in the EU 11% (Lundin et al. 2004; Fytili and Zabaniotou 2008; Hong et al. 2009). There are probably as many ways to burn sludge as there are to treat the sludge. Sludge can be incinerated in dedicated incineration plants, or it can be used as a fuel in energy production or industrial production.

The advantages of incineration are:

A large reduction in sludge volume—approximately 70% of the dry matter in the sludge disappears in incineration

Thermal destruction of toxic organic compounds

Recovering energy content of sludge

The environmental drawback of incineration is that it produces emissions: emissions to the air of particles, acid gases, greenhouse gases, and volatile organic compounds (VOCs); ashes and flue gas treatment residues, which must be disposed of to landfills; and water emissions from flue gas treatment wet processes. For WWTP owners, incineration involves high capital costs for both the incineration equipment and emissions treatment, to avoid air pollution.

Sludge incineration is a complex process, and the amounts and nature of the combustion products, and whether the incineration is an energy source or an energy sink, depend on a number of factors (Khiari et al. 2004; Dennis et al. 2005; Nowicki et al. 2011):

Effluent nature: Percentage organic matter, calorific content, percentage dry matter, pH.

Incineration process: Temperature, pressure, residence time in reactor, heat transfer speed.

As mentioned previously, it is estimated that 25% of sewage sludge is incinerated in the United States. One reason may be that facilities that produce energy from the incineration of municipal solid waste can receive a tax credit (the Renewable Electricity Production Tax Credit) to encourage energy production while managing waste (Sun 2011).

The amount of energy contained in sewage sludge is not inconsiderable, as Tables 10.3 and 10.4 show. Fytili and Zabaniotou (2008) note that sewage sludge has a calorific value almost equal to that of brown coal. As would be expected, both tables indicate a significant decrease in the calorific value of digested sludge.

The amount of water in the sludge is critical. Khiari et al. (2004) have found that with humid sludge—10 to 30% dry matter—the amount of energy released by combustion of the organic matter does not offset the energy needed to drive off the water.

A North Carolina study comparing the emissions reported from two sludge-incinerating WWTPs and from two WWTPs that do not incinerate

TABLE 10.3

Typical Heating Values for Several Types of Sewage Sludge

Type of Sludge	Heating Value (MJ/kg of DS), Range	Heating Value (MJ/kg of DS), Typical
Raw sludge	23–29	25.5
Activated sludge	16–23	21
Anaerobically digested primary sludge	9–13	11
Raw chemically precipitated primary sludge	14–18	16
Biological filter sludge	16–23	19.5

Source: Fytili and Zabaniotou, *Renewable and Sustainable Energy Reviews*, 12(1), 116–140, 2008.

TABLE 10.4

Typical Calorific Values of Different Sludge Types

Type of Sludge	Calorific Value (kWh/t DS)
Primary sludge with physical/chemical treatment or high pollution load (>0.5 kg BOD5/kg sludge/day)	4200
Secondary sludge; biological sludge—low load (between 0.07 and 0.2 BOD5/kg sludge/day)	4100
Secondary sludge from clarified water; biological sludge—low and middle load (between 0.2 and 0.5 BOD5/kg sludge/day)	4800
Mixed sludge—mixture of primary sludge and secondary sludge from clarified water	4600
Digested sludge	3000

Source: EC, *Disposal and Recycling Routes for Sewage Sludge: Part 3: Scientific and Technical Report*, European Commission, DG Environment—B/2, Office for Official Publications of the European Communities, Luxembourg, 2001.

TABLE 10.5

Incinerator Stack Exit Gas Emissions Reported for Four North Carolina WWTPs

Site	Method	Polycyclic Organic Matter (Naphthalene) in Actual Emissions Reported, lb
Graham WWTP (2008)	Does not incinerate sludge	0
East Burlington WWTP (2007)	Does not incinerate sludge	0
High Point—Eastside WWTP (2008)	Fluidized bed sewage sludge incinerator Control system is a Venturi impingement tray scrubber 2500 dry lb/h max rate; 24 h/day, 5 days/ week	878
Rocky River WWTP (2008)	Multiple hearth sewage sludge incinerator Conditioning system is a spray quencher and two-tray impingement cooler, and a wet electrostatic precipitator 5543.5 tons dry sludge processed over 4702 h.	99.8

Source: Data from Sun, Examination of the Human Health Issues of Sewage Sludge and Municipal Solid Waste Incineration in North Carolina, Blue Ridge Environmental Defense League, 2011, http://bredl.org.

found that the incinerating WWTPs emitted significant amounts of PAHs, on the order of 100 to 800 tons per year. Interestingly, only one PAH— naphthalene—accounted for the entire amount in both cases (Sun 2011). The results are summarized in Table 10.5.

Mininni et al. (2004) have also reported on PAHs generated during incineration, and linked this to the afterburning temperature.

10.3.3 Agricultural Use (Amended Soils)

Spreading sewage sludge on agricultural land, also known as soil amendment, is attractive for a number of reasons:

Soil fertility can be improved by the nutrients, such as nitrogen, phosphorus, and potassium (N, P, K), in sludge.

Reduced amounts of artificial fertilizer are needed, saving phosphate minerals.

Soil structure can be improved by incorporating organic content (Schowanek et al. 2004).

Table 10.6 shows typical nutrient (N, P, K) amounts for several types of sewage sludge.

TABLE 10.6

Typical Nitrogen, Phosphorus, and Potassium Values for Several Types of Sewage Sludge

Type of Sludge	Nitrogen, % Volatile Matter	Phosphorus, % Dry Matter	Potassium, % Dry Matter
Primary sludge with physical/chemical treatment or high pollution load (>0.5 kg BOD5/kg sludge/day)	4.5	2	0.3
Secondary sludge; biological sludge—low load (between 0.07 and 0.2 BOD5/kg sludge/day)	7.5	2	0.3
Secondary sludge from clarified water; biological sludge—low and middle load (between 0.2 and 0.5 BOD5/kg sludge/day)	6.3	2	0.3
Mixed sludge—mixture of primary sludge and secondary sludge from clarified water	7.1	2	0.3
Digested sludge	6.2	2	0.3

Source: EC, *Disposal and Recycling Routes for Sewage Sludge: Part 3: Scientific and Technical Report,* European Commission, DG Environment—B/2, Office for Official Publications of the European Communities, Luxembourg, 2001.

However, there are some concerns about the persistence of PAHs in amended soils, and about bioaccumulation. It also has calendar limitations: in agriculture, soil is fertilized once or twice a year, but WWTPs produce sludge year-round.

10.3.3.1 Persistence of PAHs in Amended Soils

Lichtfouse et al. (2005) have studied agricultural soils repeatedly amended with sewage sludge. The crop soils they studied were amended with 1000 tons dry weight of sewage sludge per 10,000 m^2 from 1974 to 1992; further sludge additions took place from 1993 to 1999. They did not find that PAHs migrated horizontally. However, they did see persistence of PAHs in the amended soils: 7 years after sludge addition, the amended soils still had more than twice the levels of control soils. Their conclusion is that sludge PAHs are preserved in crop soils for long periods of time. Wild et al. (1990) have reported the same trend; they examined samples collected and archived between 1942 and 1984 from a control site (no soil amendment) and from an agricultural site whose soil had been amended with sewage sludge 25 times between 1942 and 1961. The soil PAH concentrations increased during the years 1942–1960, and then showed a steady decline; however, in 1984, 23 years after the last application of sewage sludge, the sludge-amended site still contained over three times as much PAHs as the control soil.

Vikelsoe et al. (2002) have reported similar results when studying the persistence of PAHs in amended soils. Their results are summarized in Table 10.7.

TABLE 10.7

Persistence of PAHs in Amended Soils

Site of Soil Sampling	Amended with Sludge?	Sum of PAH Concentration, µg/kg dw
Ejby	No. Nature preserve, with no cultivation, fertilization, or soil dressing for over 50 years.	47
Sundbylille	Yes, low amounts. Cultivated area receiving low amount of sludge: about 0.7 t dw/ha/year.	46
Bistrup (1st sampling)	Yes. 1965–1990 received about 17 t dw/ha/year. In 1990 switched to artificial fertilizer. Area used for grazing cattle.	2007
Bistrup (2 years later)	Yes. 1965–1990 received about 17 t dw/ha/year. In 1990 switched to artificial fertilizer. Area used for grazing cattle.	2520

Source: Data from Vikelsoe et al., *Persistent Organic Pollutants in Soil, Sludge and Sediment: A Multianalytical Field Study of Selected Organic Chlorinated and Brominated Compounds,* NERI Technical Report 402, National Environmental Research Institute, Denmark, 2002.

Chen et al. (2005) also report accumulation of PAHs in soils, especially high molecular weight (HMW) PAHs. They examined agricultural soils that had been irrigated with sewage wastewater after the water underwent biological treatment.

Soils that are heavily contaminated with PAHs can present health risks, even if they are not used for agriculture. Yuan et al. (2014) examined the persistent organic pollutants (POPs) in urban renewal areas in southeast Beijing. These sites had formerly been chemical-industrial sites. Along with other POPs, they found heavy loadings of PAHs, in the range of 127 to 365,926 ng/g. They noted that the total lifetime carcinogenic risks (TLCRs) at several sites were higher than the acceptable levels for recreation exposure or residential cancer risk.

10.3.3.2 Are PAHs in Sludge Regulated?

At the present time, in the United States, Canada, the EU, and many other countries, regulations dealing with sewage sludge composition when used in soil amendment are concerned with heavy metals or pathogens.

10.3.4 Landfill

Disposing of sewage sludge by landfill is currently a major option; however, as pressures increase to find environmentally responsible disposal methods, this option may be expected to decrease in the future.

TABLE 10.8

Cost of Municipal Sludge Disposal Methods in the EU

Method of Disposal	Price/t of Dry Matter (€, euro)
Land spreading of solid and semisolid sludge	126–185
Landfilling	309
Co-incineration	332
Monoincineration	411

Source: EC, *Environmental, Economic and Social Impacts of the Use of Sewage Sludge on Land: Summary Report 1: Assessment of Existing Knowledge*, Contract DG ENV.G.4/ETU/2008/0076r, European Commission, DG Environment, Brussels, Belgium, 2010.

Kalmykova et al. (2013) have demonstrated that landfill leachate contains significant amounts of PAHs, higher, in fact, than expected for such hydrophobic compounds, which hints that the mobility of these pollutants may be underestimated. The amount of PAHs seen in landfill leachate and storm water from waste-sorting sites was an order of magnitude higher than that found in urban storm water.

Landfilling is used because it is cheaper than incineration; however, it is approximately twice as expensive as soil amendment. Table 10.8 gives estimated costs for sewage sludge disposal in the EU in the first decade of the 21st century.

10.4 PAHs in Sediments

Most PAHs are thought to leave the WWTP in the sludge, but a significant amount is still present in the final effluent. These are discharged, along with the treated effluent, into surface waters.

The fate of PAHs, once they are discharged in the effluent from WWTPs, is variable. In Canada, Yunker et al. (2000) studied PAH discharges into the Fraser River from a WWTP. While PAHs were detected in the WWTP effluent, they apparently biodegraded in the water column or at the sediment-water interface in the Strait of Georgia.

On the other hand, Hung et al. (2014) has examined sludge dredged from the Kim Nguu River in Hanoi, Vietnam, and found very high values: 218,000 to 751,000 µg/kg dried weight.

There are many studies reporting the detection and quantification of PAHs in sediments; a number of them are summarized very briefly in Table 10.9.

It is no coincidence that of the five most contaminated sites in Table 10.9, four are harbors (Boston, Toulon, San Diego, and Sydney). The aerial

TABLE 10.9

PAH Levels in Sediments

Location	Total PAHs in Sediment, μg/kg Dried Weight	Notes	Reference
Boston Harbor (United States)	7,300–358,000		Wang et al. 2001
Boston Harbor (United States)	718,000	2 WWTPs were discharging 500,000 kg dw hydrocarbons per year into the harbor, at this period	Shiaris and Jambard-Sweet 1986
Lake Michigan (United States)	213–1291	PAH levels showed a 2 to 10 times decrease over 20–30 years	Huang et al. 2014
Green Bay, Wisconsin (United States)	0.460–8.04 ppm		Su et al. 1998
Penobscot Bay, Maine (United States)	286–8794 ppb		Johnson et al. 1985
Susquehanna River, Chesapeake Bay watershed (United States)	1547–9847		Ko et al. 2007
San Diego Bay, California (United States)	108,000–120,000	Values are normalized for the TOC measurements because the Bay is heavily polluted with metals (Cd, Cr, Cu, Ni, Pb, Zn)	Zeng and Vista 1997
Toulon Harbor, France	Up to 48,000		Benlahcen et al. 1997
Roskilde Bay, Denmark	438–2581	Sediment close to Roskilde harbor most affected	Vikelsoe et al. 2002
Sydney Harbor, Australia	<100–380,000		McCready et al. 2000

(Continued)

TABLE 10.9 (*Continued*)

PAH Levels in Sediments

Location	Total PAHs in Sediment, μg/kg Dried Weight	Notes	Reference
Kim Nguu River, Hanoi (Vietnam)	218,000–751,000	Minimal treatment of sludge at WWTPs	Hung et al. 2014
Pearl River Estuary, Macao, China	4017–12,741, port interior; 294–1989, rural inshore		Mai et al. 2003
Klang Strait (Malaysia)	994	Not WWTP sludge	Sany et al. 2014
Malacca River and Prai River (Malaysia)	716–1210 (Malacca River); 1102–7938 (Prai River)	Petrogenic and pyrogenic sources	Keshavarzifard et al. 2014
Langat Estuary, Selangor state (Malaysia)	322–2480		Zakaria and Mahat 2006
River Plate Estuary (Argentina)	3–2100	Sediments close to sewers and port most affected	Colombo et al. 2006
Iguacu River Basin, Curitiba (Brazil)	131–1713	9 sites; industrial and urban areas	Leite et al. 2011
Abu Qir Bay (Egypt)	Sediment: MDL[a] to 2660; mussels: 24–3880	Not WWTP sludge; mussels showed higher PAH concentrations than surrounding sediments	Khairy et al. 2009
Hormuz Strait, Persian Gulf	72–278 (avg. 131)	Considered low to moderately polluted with PAHs	Rahmanpoor et al. 2014
Chao Phraya River, estuary, canals, and coast (Thailand)	2290 (canals), 263 (river), 179 (estuary), 50 (coastal areas)	Considered low to moderately polluted with PAHs	Boonyatumanond et al. 2006

[a] MDL, method detection limit.

deposits of an entire watershed are washed into harbors by rainwater, via sewer systems, WWTPs, and surface waters. If the harbor, or the watershed associated with it, happens to include a large urban area, then it can receive huge amounts of PAHs. Thus, Massachusetts Bay receives an estimated 640 kg/year of PAHs (Menzie et al. 2002); the Southern California Bight, 730 kg/year (Steinberger and Schiff 2003); and Narragansett Bay (between Rhode Island and Massachusetts), 680 kg/year (Hoffman et al. 1984).

There are many factors of course: size of the watershed in square kilometers, population equivalents living in the watershed, frequency and amount of precipitation, and length of dry season, to name a few (Stein et al. 2006). River size and speed are also factors: Ko et al. (2007) have reported that PAH levels in Chesapeake Bay sediment were related to flow rates of the Susquehanna River. Also, the characteristics of the water column are critical: for example, green, blue, and yellow algae have different effectivities in PAH removal.

It is often pointed out that when an aqueous environment receives a single huge injection of organic pollutants (e.g., an oil spill), the environment, if free of further doses of organic pollutants, seems to be capable of recovering. It may therefore seem odd that some harbor environments, with their much lower doses, cannot free themselves of PAHs in a similar manner. It may be that sediments in general contain an appropriate flora of PAH-degrading microorganisms, but in very small amounts. The environment is anaerobic, and anaerobic degradation is known to be slow (Dua et al. 2002; Haritash and Kaushik 2009). A site may recover, albeit slowly, from a single huge injection of PAHs, such as an oil spill, if given enough time and no further injections. Other sites, such as harbors, receiving much smaller continuous doses, do not recover, but instead accumulate PAHs. The anaerobic breakdown is slower than the continuous influx, and the PAHs inexorably accrue.

How do we put the measurements in Table 10.9 into perspective? A classification system that is often used is shown in Table 10.10 (Johnson et al. 1985; Benlahcen et al. 1997; Leite et al. 2011).

TABLE 10.10

Classification System for PAHs in Sediments

PAH Concentration, µg/kg	Contamination Level
Above 500	Highly contaminated
250–500	Moderately contaminated
Below 250	Slightly contaminated

Source: Johnson et al., *Marine Environmental Research*, 15(1), 1–16, 1985; Benlahcen et al., *Marine Pollution Bulletin*, 34(5), 298–305, 1997; Leite et al., *Journal of Environmental Sciences*, 23(6), 904–911, 2011.

At first glance, the divisions may seem far too low; most of the sites studied would be classed as highly contaminated. However, it should be noted that the PAH contents of plants and animals may be much higher than the PAH contents of the soil or water in which they live (ATSDR 1996). Due to accumulation along the chain, a seemingly small amount of PAHs in the source can result in high amounts at the other end of the food chain. Vikelsoe et al. (2002) have reported a buildup of PAH concentration—higher concentration in the environment than in the source—in sediment. Khairy et al. (2009) have reported higher concentrations of PAHs in mussels than was found in the surrounding sediments. In general, bioconcentration factors vary between 10 and 10,000, depending on species and type of PAH (EFSA 2008).

More study is needed in this area. Why the alarming accumulation in the Kim Nguu River, but biodegradation in the Fraser River? Which compounds are degraded (or volatized) quickly in freshwater and seawater, and which are persistent enough to accumulate and possibly be ingested? What role does the loading play? Much lower PAH loads were seen in the Fraser River study. From what PAH loadings can the environment recover? What PAHs, and at what doses, are phytotoxic to algae? What happens, to put it bluntly, between the WWTP discharge pipe and the sediments?

10.5 Factors Affecting PAH Degradation in Soils and Sediments

In soil and sediment environments, microbial metabolism by bacteria, fungi, and algae is the major degradation route for PAHs. Photolysis, hydrolysis, and oxidation are not considered to be important degradation processes, especially as the number of rings increases (Park et al. 1990; Wild and Jones 1993; ATSDR 1995).

There are many excellent reviews of this subject. We will not attempt to duplicate that work here. Instead, for more information, the reader is referred to the following: Juhasz and Naidu (2000), Dua et al. (2002), Haritash and Kaushik (2009), Kanaly and Harayama (2000), and Volkering et al. (1997).

Bioremediation, or biotransformation, is the process of using microorganisms to break down or remove hazardous waste compound from the environment (Dua et al. 2002). The PAH biotransformation process involves:

Breaking down the organic compounds into less complex metabolites

Mineralization, or the complete breakdown to the end products' inorganic minerals, water, and carbon dioxide (if aerobic) or methane (if anaerobic) (Dua et al. 2002, Haritash and Kaushik 2009)

Low molecular weight PAHs tend to biodegrade more, while high molecular weight PAHs tend to bioaccumulate more (Cerniglia 1992; NVV 2003; Haritash and Kaushik 2009). In recent years, significant advances have been made in biodegradation of high molecular weight PAHs (Juhasz and Naidu 2000; Kanaly and Harayama 2000; Larsen et al. 2009).

In addition to the physical and chemical properties of the PAHs themselves, there are many environmental and soil factors affecting whether the PAHs are biodegraded, to what extent, and how quickly:

Soil characteristics: Type, structure, and particle size, moisture, nutrients, organic content, pH.

Microbial population in the soil or sediment.

Contamination history of the soil/sediment (because it can determine whether appreciable quantities of a PAH-degrading microorganism are already in place) (Herbes and Schwall 1978; Andersson and Nilsson 1999).

Presence of co-contaminants, such as heavy metals, which are toxic to the microorganisms (Bossert and Bartha 1986).

Environmental: Temperature, precipitation, oxygen concentration.

The sorption of PAHs to organic matter and particulates in the soil can also influence whether biodegradation occurs and how quickly. When PAHs undergo sorption by organic matter in the soil or sediment, their bioavailablity decreases—and therefore their potential for biotransformation is lowered. This can hamper the biodegradation of PAHs that would otherwise be expected to rapidly be metabolized (Manilal and Alexander 1991; ATSDR 1995; Kim et al. 2001). Many studies have shown that bioavailability of the PAHs can be increased by adding synthetic surfactants, or bacteria which produce biosurfactants. Volkering (1997) has pointed out that a fine line must be walked: surfactants can improve bioavailability by moving adsorbed PAHs into the aqueous phase, or they can prove toxic to crucial elements of the microbial community. The efficacy of various surfactants, their toxicity, and their fate in the environment are areas that deserve further study (Manilal and Alexander 1991; Volkering 1997; Kim et al. 2001; Haritash and Kaushik 2009).

The microbial population in the soil/sediment is an unpredictable subject; changes can occur rapidly and with seemingly little warning. It has been pointed out that the microbial community is still a black box to us (Dua et al. 2002; Haritash and Kaushik 2009).

There are many studies examining the biodegradation half-life values of various PAHs in soils and sediments. In general, the PAHs with three or fewer rings tend to have biodegradation half-life values ranging from days to a few months; PAHs with more rings have biodegradation half-life values ranging from a month to years.

10.6 Where This Is Taking Us

PAHs are organic compounds and as such can be expected to respond to biodegradation. For bioremediation to fulfill its promise as an important tool for cleaning up PAH-contaminated soils, a greater understanding of the processes involved is required. We especially need more knowledge of the factors limiting the degradation of high molecular weight PAHs (Juhasz and Naidu 2000).

As with all biological processes, there are many factors that determine fates and rates of biotransformation of PAHs. For biotransformation at the WWTP, a better understanding of the fates of PAHs can lead to optimizing the operational parameters of the digestion processes, such as temperature, hydraulic retention time, pretreatment, and so on. In this way, we can hope to see significant improvements in sludge quality.

For sewage sludges destined for agricultural use, a better knowledge of the effects of PAHs on soil fertility, soil microbial population, and bioaccumulation through the food chain is necessary for establishing relevant limits of contaminants (Schowanek et al. 2004).

For future green technologies, we need scientific knowledge of the processes of PAH degradation and especially biodegradation. Promising work is being done on many fronts, including:

Genetically tailored microbes, to improve the range of PAHs that can be biodegraded.

In situ bioremediation, using biosurfactants (bacteria-produced surfactants) tailored for the PAH profile in the contaminated soil/sediment.

Phytoremediation: Using green plants to remove PAHs or render them harmless.

Wetlands to remove PAHs from wastewater: Aquatic weeds have already been used in wetlands to treat phenanthrene (Haritash and Kaushik 2009). Plants that can treat a variety of organic pollutants would be extremely useful.

Combinations of green technologies, e.g., phytoremediation plus degradation by microbe, to maximize pollutant removal rates.

References

Alhafez, L., Muntean, N., Muntean, E., Duda, M., and Ristoiu, D. (2013). Urban sludges utilization in agriculture: Possible limitations due to their contamination with polycyclic aromatic hydrocarbons. *ProEnvironment/ ProMediu*, 6(14).

Andersson, P.-G., and Nilsson, P. (1999). *Slamspridning på åkermark. Fältförsök med kommunalt avloppsslam från Malmö och Lund under åren 1981–1997* (Field trials with municipal sewage sludge from Malmö and Lund during 1981–1997) (in Swedish). VA-Forsk Rapport 1999–22. VAV, Stockholm.

ATSDR (1995). Toxicological profile for polycyclic aromatic hydrocarbons. U.S. Department of Health and Human Services, Agency for Toxic Substances and Disease Registry, Atlanta, GA.

ATSDR (1996). Polycyclic aromatic hydrocarbons (PAHs) ToxFAQ. U.S. Department of Health and Human Services, Agency for Toxic Substances and Disease Registry, Atlanta, GA. http://www.atsdr.cdc.gov/toxfaq.html.

Augulyte, L. (2001). Dewatering technologies, handling problems and effect to the environment from municipal sewage sludge. Project report. Environmental Chemistry, Department of Chemistry, Umeå University, Umeå, Sweden.

Benlahcen, K.T., Chaoui, A., Budzinski, H., Bellocq, J., and Garrigues, Ph. (1997). Distribution and sources of polycyclic aromatic hydrocarbons in some Mediterranean coastal sediments. *Marine Pollution Bulletin*, 34(5), 298–305.

Blanchard, M., Teil, M.J., Ollivon, D., Legenti, L., and Chevreuil, M. (2004). Polycyclic aromatic hydrocarbons and polychlorobiphenyls in wastewaters and sewage sludges from the Paris area (France). *Environmental Research*, 95(2), 184–197.

Boonyatumanond, R., Wattayakorn, G., Togo, A., and Takada, H. (2006). Distribution and origins of polycyclic aromatic hydrocarbons (PAHs) in riverine, estuarine, and marine sediments in Thailand. *Marine Pollution Bulletin*, 52(8), 942–956.

Bossert, I.D., and Bartha, R. (1986). Structure-biodegradability relationships of polycyclic aromatic hydrocarbons in soil. *Bulletin of Environmental Contamination and Toxicology*, 37(1), 490–495.

Busetti, F., Heitz, A., Cuomo, M., Badoer, S., and Traverso, P. (2006). Determination of sixteen polycyclic aromatic hydrocarbons in aqueous and solid samples from an Italian wastewater treatment plant. *Journal of Chromatography A*, 1102(1), 104–115.

Cai, Q.Y., Mo, C.H., Wu, Q.T., Zeng, Q.Y., and Katsoyiannis, A. (2007). Occurrence of organic contaminants in sewage sludges from eleven wastewater treatment plants, China. *Chemosphere*, 68(9), 1751–1762.

Cerniglia, C.E. (1992). Biodegradation of polycyclic aromatic hydrocarbons. *Biodegradation*, 3(2–3), 351–368.

Chen, Y., Wang, C., and Wang, Z. (2005). Residues and source identification of persistent organic pollutants in farmland soils irrigated by effluents from biological treatment plants. *Environment International*, 31(6), 778–783.

Colombo, J.C., Cappelletti, N., Lasci, J., Migoya, M.C., Speranza, E., and Skorupka, C.N. (2006). Sources, vertical fluxes, and equivalent toxicity of aromatic hydrocarbons in coastal sediments of the Rio de la Plata Estuary, Argentina. *Environmental Science and Technology*, 40(3), 734–740.

Dai, J., Xu, M., Chen, J., Yang, X., and Ke, Z. (2007). PCDD/F, PAH and heavy metals in the sewage sludge from six wastewater treatment plants in Beijing, China. *Chemosphere*, 66(2), 353–361.

Davis, R.D. (1996). The impact of EU and UK environmental pressures on the future of sludge treatment and disposal. *Water and Environment Journal*, 10(1), 65–69.

Dennis, J.S., Lambert, R.J., Milne, A.J., Scott, S.A., and Hayhurst, A.N. (2005). The kinetics of combustion of chars derived from sewage sludge. *Fuel*, 84(2), 117–126.

Dua, M., Singh, A., Sethunathan, N., and Johri, A. (2002). Biotechnology and bioremediation: Successes and limitations. *Applied Microbiology and Biotechnology*, 59(2–3), 143–152.

EC. (2001). *Disposal and recycling routes for sewage sludge: Part 3: Scientific and technical report. European Commission*, DG Environment—B/2, Office for Official Publications of the European Communities, Luxembourg.

EC. (2010). *Environmental, economic and social impacts of the use of sewage sludge on land: Summary report 1: Assessment of existing knowledge.* Contract DG ENV.G.4/ETU/2008/0076r. European Commission, DG Environment, Brussels, Belgium.

EFSA (European Food Safety Authority). (2008). Polycyclic aromatic hydrocarbons in food: Scientific opinion of the Panel on Contaminants in the Food Chain (question no. EFSA-Q-2007–136). *The EFSA Journal*, 724, 1–114.

Fytili, D., and Zabaniotou, A. (2008). Utilization of sewage sludge in EU application of old and new methods—A review. *Renewable and Sustainable Energy Reviews*, 12(1), 116–140.

Haritash, A.K., and Kaushik, C.P. (2009). Biodegradation aspects of polycyclic aromatic hydrocarbons (PAHs): A review. *Journal of Hazardous Materials*, 169(1), 1–15.

Herbes, S.E., and Schwall, L.R. (1978). Microbial transformation of polycyclic aromatic hydrocarbons in pristine and petroleum-contaminated sediments. *Applied and Environmental Microbiology*, 35(2), 306–316.

Hoffman, E.J., Mills, G.L., Latimer, J.S., and Quinn, J.G. (1984). Urban runoff as a source of polycyclic aromatic hydrocarbons to coastal waters. *Environmental Science and Technology*, 18(8), 580–587.

Hong, J., Hong, J., Otaki, M., and Jolliet, O. (2009). Environmental and economic life cycle assessment for sewage sludge treatment processes in Japan. *Waste Management*, 29(2), 696–703.

Huang, L., Chernyak, S.M., and Batterman, S.A. (2014). PAHs (polycyclic aromatic hydrocarbons), nitro-PAHs, and hopane and sterane biomarkers in sediments of southern Lake Michigan, USA. *Science of the Total Environment*, 487, 173–186.

Hung, C.V., Cam, B.D., Mai, P.T.N., and Dzung, B.Q. (2014). Heavy metals and polycyclic aromatic hydrocarbons in municipal sewage sludge from a river in highly urbanized metropolitan area in Hanoi, Vietnam: Levels, accumulation pattern and assessment of land application. *Environmental Geochemistry and Health*, 1–14.

Johnson, A.C., Larsen, P.F., Gadbois, D.F., and Humason, A.W. (1985). The distribution of polycyclic aromatic hydrocarbons in the surficial sediments of Penobscot Bay (Maine, USA) in relation to possible sources and to other sites worldwide. *Marine Environmental Research*, 15(1), 1–16.

Ju, J.H., Lee, I.S., Sim, W.J., Eun, H., and Oh, J.E. (2009). Analysis and evaluation of chlorinated persistent organic compounds and PAHs in sludge in Korea. *Chemosphere*, 74(3), 441–447.

Juhasz, A.L., and Naidu, R. (2000). Bioremediation of high molecular weight polycyclic aromatic hydrocarbons: A review of the microbial degradation of benzo[a]pyrene. *International Biodeterioration and Biodegradation*, 45(1), 57–88.

Kalmykova, Y., Björklund, K., Strömvall, A.M., and Blom, L. (2013). Partitioning of polycyclic aromatic hydrocarbons, alkylphenols, bisphenol A and phthalates in landfill leachates and stormwater. *Water Research*, 47(3), 1317–1328.

Kanaly, R.A., and Harayama, S. (2000). Biodegradation of high-molecular-weight polycyclic aromatic hydrocarbons by bacteria. *Journal of Bacteriology*, 182(8), 2059–2067.

Keshavarzifard, M., et al. (2014). Baseline distributions and sources of polycyclic aromatic hydrocarbons (PAHs) in the surface sediments from the Prai and Malacca Rivers, Peninsular Malaysia. *Marine Pollution Bulletin*, 88(1–2), 366–372.

Khadhar, S., Higashi, T., Hamdi, H., Matsuyama, S., and Charef, A. (2010). Distribution of 16 EPA-priority polycyclic aromatic hydrocarbons (PAHs) in sludges collected from nine Tunisian wastewater treatment plants. *Journal of Hazardous Materials*, 183, 98–102.

Khairy, M.A., Kolb, M., Mostafa, A.R., El-Fiky, A., and Bahadir, M. (2009). Risk assessment of polycyclic aromatic hydrocarbons in a Mediterranean semi-enclosed basin affected by human activities (Abu Qir Bay, Egypt). *Journal of Hazardous Materials*, 170(1), 389–397.

Khiari, B., Marias, F., Zagrouba, F., and Vaxelaire, J. (2004). Analytical study of the pyrolysis process in a wastewater treatment pilot station. *Desalination*, 167, 39–47.

Kim, I.S., Park, J.S., and Kim, K.W. (2001). Enhanced biodegradation of polycyclic aromatic hydrocarbons using nonionic surfactants in soil slurry. *Applied Geochemistry*, 16(11), 1419–1428.

Ko, F.C., Baker, J., Fang, M.D., and Lee, C.L. (2007). Composition and distribution of polycyclic aromatic hydrocarbons in the surface sediments from the Susquehanna River. *Chemosphere*, 66(2), 277–285.

Larsen, S.B., Karakashev, D., Angelidaki, I., and Schmidt, J.E. (2009). Ex-situ bioremediation of polycyclic aromatic hydrocarbons in sewage sludge. *Journal of Hazardous Materials*, 164(2), 1568–1572.

Leite, N.F., Peralta-Zamora, P., and Grassi, M.T. (2011). Distribution and origin of polycyclic aromatic hydrocarbons in surface sediments from an urban river basin at the metropolitan region of Curitiba, Brazil. *Journal of Environmental Sciences*, 23(6), 904–911.

Lichtfouse, E., Sappin-Didier, V., Denaix, L., Caria, G., Metzger, L., Amellal-Nassr, N., and Schmidt, J. (2005). A 25-year record of polycyclic aromatic hydrocarbons in soils amended with sewage sludges. *Environmental Chemistry Letters*, 3(3), 140.

Lundin, M., Olofsson, M., Pettersson, G.J., and Zetterlund, H. (2004). Environmental and economic assessment of sewage sludge handling options. *Resources, Conservation and Recycling*, 41(4), 255–278.

Mai, B., et al. (2003). Distribution of polycyclic aromatic hydrocarbons in the coastal region off Macao, China: Assessment of input sources and transport pathways using compositional analysis. *Environmental Science and Technology*, 37(21), 4855–4863.

Mailler, R., et al. (2014). Priority and emerging pollutants in sewage sludge and fate during sludge treatment. *Waste Management*, 34(7), 1217–1226.

Manilal, V.B., and Alexander, M. (1991). Factors affecting the microbial degradation of phenanthrene in soil. *Applied Microbiology and Biotechnology*, 35(3), 401–405.

McCready, S., Slee, D.J., Birch, G.F., and Taylor, S.E. (2000). The distribution of polycyclic aromatic hydrocarbons in surficial sediments of Sydney Harbour, Australia. *Marine Pollution Bulletin*, 40(11), 999–1006.

Menzie, C.A., Hoeppner, S.S., Cura, J.J., Freshman, J.S., and LaFrey, E.N. (2002). Urban and suburban storm water runoff as a source of polycyclic aromatic hydrocarbons (PAHs) to Massachusetts estuarine and coastal environments. *Estuaries*, 25(2), 165–176.

Mininni, G., Sbrilli, A., Guerriero, E., and Rotatori, M. (2004). Polycyclic aromatic hydrocarbons formation in sludge incineration by fluidised bed and rotary kiln furnace. *Water, Air and Soil Pollution*, 154(1–4), 3.

Nowicki, L., Antecka, A., Bedyk, T., Stolarek, P., and Ledakowicz, S. (2011). The kinetics of gasification of char derived from sewage sludge. *Journal of Thermal Analysis and Calorimetry*, 104(2), 693–700.

NVV. (2003). *Organic contaminants in sewage sludge: Review of studies regarding occurrence and risks in relation to the application of sewage sludge in agricultural soil*. NVV Rapport 5217. Naturvårdsverket (Swedish Environmental Protection Agency), Stockholm.

Oleszczuk, P. (2007). Changes of polycyclic aromatic hydrocarbons during composting of sewage sludges with chosen physico-chemical properties and PAHs content. *Chemosphere*, 67, 582–591.

Park, K.S., Sims, R.C., Dupont, R.R., Doucette, W.J., and Matthews, J.E. (1990). Fate of PAH compounds in two soil types: Influence of volatilization, abiotic loss and biological activity. *Environmental Toxicology and Chemistry*, 9(2), 187–195.

Pérez, S., Guillamón, M., and Barceló, D. (2001). Quantitative analysis of polycyclic aromatic hydrocarbons in sewage sludge from wastewater treatment plants. *Journal of Chromatography A*, 938(1), 57–65.

Rahmanpoor, S., Ghafourian, H., Hashtroudi, S.M., and Bastami, K.D. (2014). Distribution and sources of polycyclic aromatic hydrocarbons in surface sediments of the Hormuz strait, Persian Gulf. *Marine Pollution Bulletin*, 78(1), 224–229.

Sany, S.B.T., Hashim, R., Salleh, A., Rezayi, M., Mehdinia, A., and Safari, O. (2014). Polycyclic aromatic hydrocarbons in coastal sediment of Klang Strait, Malaysia: Distribution pattern, risk assessment and sources. *PloS One*, 9(4), e94907.

Schowanek, D., Carr, R., David, H., Douben, P., Hall, J., Kirchmann, H., Patria, L., Sequi, P., Smith, S., and Webb, S. (2004). A risk-based methodology for deriving quality standards for organic contaminants in sewage sludge for use in agriculture—Conceptual framework. *Regulatory Toxicology and Pharmacology*, 40(3), 227–251.

Shiaris, M.P., and Jambard-Sweet, D. (1986). Polycyclic aromatic hydrocarbons in surficial sediments of Boston Harbour, Massachusetts, USA. *Marine Pollution Bulletin*, 17(10), 469–472.

Stein, E.D., Tiefenthaler, L.L., and Schiff, K. (2006). Watershed-based sources of polycyclic aromatic hydrocarbons in urban storm water. *Environmental Toxicology and Chemistry*, 25(2), 373–385.

Steinberger, A., and Schiff, K.C. (2003). Characteristics of effluents from large municipal wastewater treatment facilities between 1998 and 2000. In *Southern California Coastal Water Research Project, Annual Report 2001–2002*. Southern California Coastal Water Research Project, Costa Mesa, CA, pp. 2–12.

Su, M.C., Christensen, E.R., and Karls, J.F. (1998). Determination of PAH sources in dated sediments from Green Bay, Wisconsin, by a chemical mass balance model. *Environmental Pollution*, 99(3), 411–419.

Sun, K. (2011). Examination of the human health issues of sewage sludge and municipal solid waste incineration in North Carolina. Blue Ridge Environmental Defense League. http://bredl.org.

Torretta, V., and Katsoyiannis, A. (2013). Occurrence of polycyclic aromatic hydrocarbons in sludges from different stages of a wastewater treatment plant in Italy. *Environmental Technology*, 34(7), 937–943.

Vikelsoe, J., Thomsen, M., Carlsen, L., and Johansen, E. (2002). *Persistent organic pollutants in soil, sludge and sediment: A multianalytical field study of selected organic chlorinated and brominated compounds*. NERI Technical Report 402. National Environmental Research Institute, Denmark.

Villar, P., Callejon, M., Alonso, E., Jimenez, J.C., and Guiraum, A. (2006). Temporal evolution of polycyclic aromatic hydrocarbons (PAHs) in sludge from wastewater treatment plants: Comparison between PAHs and heavy metals. *Chemosphere*, 64, 535–541.

Volkering, F., Breure, A.M., and Rulkens, W.H. (1997). Microbiological aspects of surfactant use for biological soil remediation. *Biodegradation*, 8(6), 401–417.

Wang, X.C., Zhang, Y.X., and Chen, R.F. (2001). Distribution and partitioning of polycyclic aromatic hydrocarbons (PAHs) in different size fractions in sediments from Boston Harbor, United States. *Marine Pollution Bulletin*, 42(11), 1139–1149.

Werle, S., and Dudziak, M. (2014). Analysis of organic and inorganic contaminants in dried sewage sludge and by-products of dried sewage sludge gasification. *Energies*, 7(1), 462–476.

Werther, J., and Ogada, T. (1999). Sewage sludge combustion. *Progress in Energy and Combustion Science*, 25(1), 55–116.

Wild, S.R., and Jones, K.C. (1993). Biological and abiotic losses of polynuclear aromatic hydrocarbons (PAHs) from soils freshly amended with sewage sludge. *Environmental Toxicology and Chemistry*, 12(1), 5–12.

Wild, S.R., Waterhouse, K.S., McGrath, S.P., and Jones, K.C. (1990). Organic contaminants in an agricultural soil with a known history of sewage sludge amendments: Polynuclear aromatic hydrocarbons. *Environmental Science and Technology*, 24(11), 1706–1711.

Yuan, G.L., Wu, H.Z., Fu, S., Han, P., and Lang, X.X. (2014). Persistent organic pollutants (POPs) in the topsoil of typical urban renewal area in Beijing, China: Status, sources and potential risk. *Journal of Geochemical Exploration*, 138, 94–103.

Yunker, M.B., et al. (2000). Assessment of natural and anthropogenic hydrocarbon inputs using PAHs as tracers: The Fraser River Basin and Strait of Georgia, 1987–1997. Burrard Inlet Environmental Action Program.

Zakaria, M.P., and Mahat, A.A. (2006). Distribution of polycyclic aromatic hydrocarbon (PAHs) in sediments in the Langat Estuary. *Coastal Marine Science*, 30(1), 387–395.

Zeng, E.Y., and Vista, C.L. (1997). Organic pollutants in the coastal environment off San Diego, California. 1. Source identification and assessment by compositional indices of polycyclic aromatic hydrocarbons. *Environmental Toxicology and Chemistry*, 16(2), 179–188.

Zeng, X.Y., Cao, S.X., Zhang, D.L., Gao, S.T., Yu, Z.Q., Li, H.R., Sheng, G.-Y., and Fu, J.M. (2012). Levels and distribution of synthetic musks and polycyclic aromatic hydrocarbons in sludge collected from Guangdong Province. *Journal of Environmental Science and Health A*, 47(3), 389–397.

Index